Studies in European Economic Law and Regulation

Volume 29

The series "Studies in European Economic Law and Regulation" is devoted to the analysis of European Economic Law. The series' scope covers a broad range of topics within economics law including, but not limited to, the relationship between EU law and WTO law; free movement under EU law and its impact on fundamental rights; antitrust law; trade law; unfair competition law; financial market law; consumer law; food law; and health law. These subjects are approached both from doctrinal and interdisciplinary perspectives.

The series accepts monographs focusing on a specific topic, as well as edited collections of articles covering a specific theme or collections of articles. All contributions are subject to rigorous double-blind peer-review.

Marianna Vanuzzo

Transparency in EU Law

From Legal Principles to a Food Sector Analysis

Springer

Marianna Vanuzzo
Università della Svizzera italiana (USI)
Lugano, Switzerland

Catholic University of the Sacred Heart
Milan, Italy

ISSN 2214-2037 ISSN 2214-2045 (electronic)
Studies in European Economic Law and Regulation
ISBN 978-3-032-02344-5 ISBN 978-3-032-02345-2 (eBook)
https://doi.org/10.1007/978-3-032-02345-2

This Springer imprint is published by the registered company Springer Nature Switzerland AG
The registered company address is: Gewerbestrasse 11, 6330 Cham, Switzerland

To my family

Foreword: Transparency Legally: What's on Offer?

EU food regulation has traditionally balanced the objectives of market integration within the internal market, consumer safety, product diversity, and fair consumer choice. Recently, sustainability has been added to these goals, and there is growing interest in also including the protection of innovation. However, the most recent legal addition to this framework is the principle of transparency. Since the Transparency Regulation (EU) 2019/1381 came into force, businesses, regulatory authorities, and consumers have been working to implement it effectively. The rationale behind its introduction is that increased transparency enhances public trust, regulatory effectiveness, and the overall legitimacy of EU food law.

Transparency in EU food law is not merely a procedural requirement; it is a foundational principle that affects how laws are drafted, implemented, and enforced. In an era of heightened public scrutiny and increasing demand for accountability, the need for transparency has never been more pressing. The public, along with various stakeholders, demands to know not only that food is safe but also how that determination is made, who influences those decisions, and what scientific evidence or policy considerations underlie regulatory actions.

This book aims to provide a comprehensive analysis of transparency in EU food law, exploring its development, current state, and future prospects. It delves into the legal mechanisms designed to ensure transparency, such as the open access to documents, public consultations, and the role of scientific advisory bodies. It also critically examines the challenges and limitations inherent in the current system, including issues related to proprietary information, and the balancing of transparency with other legitimate interests, such as trade secrets and innovation.

The chapters that follow are the product of extensive research and reflection. They offer not only a detailed exposition of the legal frameworks but also engage with the broader implications of transparency for the EU's food regulatory system. These insights are particularly relevant as the EU continues to navigate the complexities of global food supply chains, the rise of new technologies, and the increasing interconnectedness of food safety with environmental and public health policies.

Ultimately, this book seeks to contribute to the ongoing debate about how best to achieve a transparent, accountable, and effective regulatory system for food in the

European Union. It is our hope that it will serve as a valuable resource for academics, policymakers, practitioners, and anyone with an interest in the governance of food law and its impact on society.

University of Bayreuth,
Bayreuth, Germany
March 2025

Kai P. Purnhagen

Acknowledgements

This book originated from a revised version of my doctoral thesis on 'Transparency in EU food law' defended on 26 May 2022 at the Catholic University in Milan.

I owe my deepest gratitude to Professor Francesco Bestagno whose teaching originally sparked my interest in EU law and whose mentorship has been constant over the years. A special word of thanks to Professor Andrea Santini, for sharing his profound expertise on transparency in EU law and his incisive comments.

This book would not have been possible without Professor Hanna Schebesta who invited me to Wageningen University & Research and guided me through new research methodologies. I extend my thanks to Professor Josephine van Zeben, for welcoming me to the LAW Group and for her generous support.

I express my gratitude to Professor Kai Purnhagen, Series editor, for believing in this project, and Anja Trautmann for her assistance. Professor Ilaria Espa deserves a special appreciation for involving me in new areas of law and policy at USI Lugano and encouraging me to bring this book to publication.

Finally, of course, thank you to my family, friends, and colleagues for their love, friendship, and support.

June 2025 Marianna Vanuzzo

Competing Interests The author has no competing interests to declare that are relevant to the content of this manuscript.

Contents

Abbreviations

AG	Advocate General
BSE	Bovine Spongiform Encephalopathy
CAQDAS	Computer-Assisted Qualitative Document Analysis Software
CCPR	International Covenant on Civil and Political Rights
CFI	Court of First Instance
CJEU	Court of Justice of the European Union
EC	European Community
ECHA	European Chemicals Agency
ECHR	European Convention of Human Rights
ECI	European Citizens' Initiative
ECtHR	European Court of Human Rights
EFSA	European Food Safety Authority
EMA	European Medicines Agency
ETI	European Transparency Initiative
EU	European Union
FBO	Food Business Operator
FIC	Food Information to Consumers
GFL	General Food Law
GMOs	Genetically Modified Organisms
IARC	International Agency for Research on Cancer
NHCR	Nutrition and Health Claims Regulation
NPBTs	New Plant Breeding Techniques
PPP	Plant Protection Products
RASFF	Rapid Alert System for Food and Feed
REFIT	Regulatory Fitness and Performance Programme
SME	Small and Medium-sized Enterprise
TEC	Treaty Establishing the European Community
TEU	Treaty on European Union
TEC	Treaty Establishing the European Community

TFEU	Treaty on the Functioning of the European Union
UNECE	United Nations Economic Commission for Europe
WTO	World Trade Organisation

Chapter 1
Introduction

Abstract This chapter explains the main purpose of the book and gives an introduction outlining the structure and the methodology used.

1.1 Why a Book on Transparency in EU Law?

1.1.1 Aim of the Book

The main purpose of this book is to examine how the topic of transparency has been addressed in the European law domain, particularly in the food law framework. This perspective is particularly important in the aftermath of the Regulation (EU) 2019/1381 on the transparency and sustainability of the EU risk assessment in the food chain, which entered into force on 27 March 2021.[1] The Regulation on transparency brings innovative changes to the General Food Law (Regulation (EC) No 178/2002 or 'GFL'), which, as known, sets common measures for the protection of human health, safety, and consumer protection through fundamental principles and legitimate purposes relating to food law. In the last few years, the lack of transparency perceived by citizens and the results of the "Fitness Check" of the General Food Law have created a growing interest in this issue.

Choosing this topic requires approaching "transparency" from a broader view. *Transparency* is a concept extremely difficult to define and to address in legal contexts.

At the European level, transparency was initially conceived as a principle of good governance more than a principle of law. Moreover, there is no shared definition of transparency. The idea of transparency itself is often linked with other

[1] Regulation (EU) 2019/1381 of the European Parliament and of the Council of 20 June 2019 on the transparency and sustainability of the EU risk assessment in the food chain and amending Regulations (EC) No 178/2002, (EC) No 1829/2003, (EC) No 1831/2003, (EC) No 2065/2003, (EC) No 1935/2004, (EC) No 1331/2008, (EC) No 1107/2009, (EU) 2015/2283 and Directive 2001/18/EC, OJ L 231, 6.9.2019, pp. 1–28.

M. Vanuzzo, *Transparency in EU Law*, Studies in European Economic Law and Regulation 29, https://doi.org/10.1007/978-3-032-02345-2_1

concepts, such as democracy and accountability. The extensive literature on transparency highlights that its meaning is unclear, and it takes on different facets depending on the field of interest. For example, a large number of governance and political studies examine the issue of transparency.[2]

Instead, the EU legal framework has only partially addressed the issue of transparency, and its value as a legal principle of EU law is not unanimously accepted. However, the focus on the EU law domain relies on the fact that EU food law is a subset of EU law. In particular, EU food law is both part and *lex specialis* in respect of EU law. As it is known, food law is multilevel, and the cornerstone topics of food safety, food security and food trade are addressed by national law, regional law and international law.[3]

The topic of transparency has been a key concern since the beginning of the European food law development. Significantly, the White Paper on Food Safety of 2000 concludes that "greater transparency at all levels of Food Safety policy is the thread running through the whole White Paper and will contribute fundamentally to enhancing consumer confidence in EU food safety policy".[4]

The General Food Law devotes two sections to the implementation of the *principle of transparency*. The first is expressed in Articles 9 and 10 and refers to consultation and information for citizens. The second mention of transparency in the General Food Law is found in Article 38, which concerns the European Food Safety Authority. Still, it seems that limiting transparency to only a few provisions of the General Food Law that explicitly mention it risks underestimating its scope. Accordingly, this book adopts a broader approach. Exploring how the principle of transparency is embedded in food law and whether it has an instrumental and unifying role is still an unexplored issue that can add value to the research.

Transparency likewise seems instrumental to the European food law's credibility and citizens' participation in the democratic order. This value is even more emphasised in the relationships among the actors of the food system. These are the ones directly or indirectly involved in the food chain, including consumers, producers, workers, employers, traders, and transporters. Recital 9 General Food Law—expressing that it "is necessary to ensure that consumers, other stakeholders and trading partners have confidence in the decision-making processes underpinning food law, its scientific basis and the structures and independence of the institutions protecting health and other interests"—refers to the prominence of a valuable relationship among all parties involved. In practice, the increasing sensitivity to direct interest representation in the field of food safety is the result of the crisis of confidence in the government of experts that followed the BSE and *dioxin affair* emergencies. Moreover, the collective approach involving public authorities, private-sector actors, NGOs, social partners, academia and citizens is emphasised in

[2] See, Gupta and Mason (2014), Chapter 1.

[3] See Van der Meulen (2011), p. 33, Schebesta and Purnhagen (2024), p. 384.

[4] White Paper on Food Safety, COM (1999) 719 final, Brussels, 12 January 2000.

documents such as the *Farm to Fork Strategy*[5] and the recent *Vision for Agriculture and Food*.[6]

The book, therefore, considers the actors involved and highlights the corresponding relevant interests. Among the stakeholders, the private sector is of particular interest. Regarding the impact of disclosure on study design and regulatory strategies, the work aims to shed light on whether the new rules on transparency could pose the grounds for addressing the shortcomings of General Food Law and what their effects on business interests could be.

1.1.2 Specific Objectives

It is essential first to clarify the objectives and the specific research questions addressed throughout the book. *First, the book aims to reach a better legal understanding of transparency in European food law.*

In this vein, it is valuable to examine whether transparency obligations are present in the food law legislation, especially in the General Food Law, even if they do not refer explicitly to transparency.

Within the complex framework of various actors involved, it is particularly interesting to consider whether transparency applies to specific actors, namely the private sector.

Therefore, *the second objective of the book aims to investigate whether transparency can be applied in or to the private sector.* Transparency obligations can fall on private bodies and they are not limited to the public authorities' duties. Accordingly, the private actors' role is key to assessing how transparency is addressed in EU law.

1.1.3 Research Questions and Structure of the Book

In order to tackle the objectives of the book, it is beneficial to clarify the *research questions* related to these targets. In its structure, the book takes into account the research questions that are useful for developing the arguments. Therefore, each chapter addresses a research question and its corresponding sub-questions, which are answered within the relevant chapter.

---

[5] Communication from the Commission to the European Parliament, the Council, the European Economic and Social Committee and the Committee of the Regions. A Farm to Fork Strategy for a fair, healthy and environmentally-friendly food system, COM/2020/381 final.

[6] Communication from the Commission to the European Parliament, the Council, the European Economic and Social Committee and the Committee of the Regions. A Vision for Agriculture and Food Shaping together an attractive farming and agri-food sector for future generations, COM/2025/75 final.

Since the first objective is to provide a better legal understanding of transparency in European food law, it is necessary to ask: *How is the principle of transparency currently enshrined in EU law and EU food law specifically and are they the same?.* This question presupposes the existence of a principle of transparency in EU law and food law specifically. This perspective becomes clear throughout the book. To address this question, the theme has been subdivided into distinct sub-research questions serving as logical premises.

Since transparency is a concept that has also emerged in legal studies, the first step is to clarify *the genesis and role of transparency in law.* In doing so, the sub-research questions that have been developed are: *What are the dimensions of transparency? What purposes does transparency serve under the dimensions of transparency? Who is the main target of transparency?*

The second step is to analyse *whether there is a general principle of transparency in the European Union law.* In particular, examining the EU legal framework and whether transparency can be regarded as a well-established principle of law is useful for clarifying the legal dimension of transparency in EU law. This point is key to understanding the general framework, which encompasses food law as a subset.

Once a clearer understanding of the EU law framework is established, the book analyses in depth *how the principle of transparency is enshrined in current food law.* To reach a partial understanding of these questions, the sub-questions are: *What are the dimensions of transparency in food law? Are there any transparency obligations arising from food law regulations, even if they do not make explicit reference to transparency? How do the general principle and the specific principle compare?*

Chapter 2 begins by illustrating the possible meanings of transparency in the literature. More specifically, it describes its origin in the legal domain, focusing on the fields of international law and EU law. Following an in-depth exploration of the notion, the book highlights that transparency can be described as a "multidimensional" concept linked to several purposes. Due to the often confusing and overlapping meanings of transparency, the chosen approach serves to clarify its purposes by describing its related "dimensions".

The chapter identifies and chooses these transparency dimensions, namely "democracy", "legitimacy", "accountability", "realisation of individual rights", and "market efficiency". Each of these dimensions can help the reader gain a deeper understanding of the concept of transparency itself. Therefore, identifying and describing these dimensions constitute the basis of the study. Due to the flexibility and vagueness of the transparency concept, the analysis explain why *information* could be the core notion behind it. Chapter 2 also reflects upon the target of transparency obligation. Often, EU administrative law scholars have limited these obligations to the public authorities' duties. Nevertheless, transparency obligations can also fall on private bodies.

Chapter 3 describes the EU law doctrinal approach to transparency and the most pivotal CJEU case law on the subject matter. The scope is to answer the sub-question concerning the existence of a general principle of transparency in EU law. As will be seen, the focus is consistently confirmed to be on access to documents rather than on transparency itself from both doctrinal and judicial points of view. Before addressing this point, the development of the principle of transparency in various fields of EU law is explained, along with the distinction between active and passive transparency. Furthermore, in EU law, transparency has emerged alongside other well-established legal principles, such as sound administration and legal certainty. Lastly, the extent to which Regulation (EC) No 1049/2001 on access to documents and Regulation (EC) No 1367/2006 remain the core legal framework for transparency in EU law is examined.

Chapter 4 proposes a commented analysis of the General Food Law, transitioning from EU law to the specific domain of EU food law. It considers the policy issues behind its adoption to facilitate a wider understanding of the EU food law legal framework. Furthermore, after exposing the GFL structure and the "Fitness Check", it focuses on the amendments introduced by Regulation (EU) 2019/1381. The analysis then deals with the existence of transparency obligations and the application of the "dimensions" framework. Moreover, Chap. 4 takes into account the judicial approach to transparency in the food sector. Following an in-depth CJEU case law analysis, the chapter draws some conclusions on the issue.

In order to tackle the second objective, the specific question is: *How is and could the principle of transparency be applied in or to the private sector?*. Due to the legal framework considered, the sub-question in the context of the new transparency rules is: *How far does the principle of transparency constrain the actions of private actors at the moment?*.

Chapter 5 describes the legal background and the specific rules on confidentiality that arise from the novelties introduced by the Transparency Regulation. The analysis suggests that the constraints and burdens imposed on private actors may not effectively incentivise innovations within the EU.

In the Conclusions, the potential future role of the principle of transparency is canvassed, and some reflections are offered.

1.2 Methodological Design

The book uses law as a primary source. The legal sources cover Treaties and regulations laid down by legislatures, judicial decisions, and juristic writing of qualified publicists as a subsidiary source. The analysis of the legislation also includes preparatory work, factsheets, and all the elements related to the adoption of a law.

The research conducted relies mostly on different ways of analysis to answer the questions posed. Firstly, the book deploys classical legal doctrinal analysis.[7] Moreover, it considers the most significant judicial decisions from the Court of Justice of the European Union. The book also addresses the legal relationships

[7] In traditional legal research, authoritative texts such as legislation, case law and doctrinal literature are considered "the main formal sources of information for understanding positive law", see Langbroek et al. (2017), p. 2.

between areas of law—such as EU law in general and food law specifically—and their hierarchical relations, as well as conflicts and legal instruments that integrate matters between these areas.

The legislative and judicial conceptions of transparency in European food law are separately compared in the book. Particularly, a conceptual analysis is undertaken in the interpretative examination of the General Food Law. More in detail, a systematic inductive approach is applied to reach a deeper understanding of the General Food Law's transparency dimensions. Meanwhile, Computer Assisted Qualitative Document Analysis Software (CAQDAS) is used to conduct empirical legal research in CJEU case law. As explained in detail in the paragraphs regarding the case-law analysis, this methodology is a useful tool to enhance the research's systematic approach and to avoid the problems of lacking robustness and cognitive overload.[8] Through the blend of classical doctrinal research, legal interpretation and innovative methodologies in law a wider and more consistent view of the subject matter can be achieved. In particular, doctrinal research focuses on legislation, case law and other legal sources. Legal interpretation and innovative tools such as CAQDAS further aim to explain the meaning of legislation and case law.

References

Gupta A, Mason M (2014) Transparency in global environmental governance: critical perspectives. MIT Press, Cambridge

Langbroek P, van den Bos K, Thomas MS et al (2017) Editorial methodology of legal research: challenges and opportunities. Utrecht Law Rev 13(3):1–8

Schebesta H (2018) Content analysis software in legal research: a proof of concept using ATLAS. Ti. Tilburg Law Rev 23(1–2):23–33

Schebesta H, Purnhagen K (2024) EU food law. Oxford University Press, Oxford

Van Der Meulen B (2011) Private food law: governing food chains through contract law, self-regulation, private standards, audits and certification schemes. Wageningen Academic Publishers, Wageningen

[8] Schebesta (2018), p. 25.

Chapter 2
The Genesis and Role of Transparency in Law

Abstract This chapter illustrates the possible meanings of transparency in the literature. It describes its origin in the legal domain, focusing on international law and EU law. Following an extensive literature review on the notion of transparency, the book highlights that it can be described as a "multidimensional" concept linked with several purposes. Due to the often overlapping meanings of transparency, the chosen approach serves to clarify its purposes by describing its related "dimensions". "Democracy", "legitimacy", "accountability", "realisation of individual rights", and "market efficiency" are the transparency dimensions identified and chosen in this book. Each of these dimensions can help readers gain a deeper understanding of the transparency concept. This chapter also reflects upon the target of transparency obligations.

2.1 Introduction

Transparency could be an overused word in the universal language as well as in the legal context. Undoubtedly, it is not just challenging to define in legal terms, but also "difficult to grasp in terms of content".[1]

No one can object that transparency fascinates everyone and is demanded by everyone.[2] The concept indeed evokes positive values inherent in society, such as democracy and public participation in decision-making. The common idea is that transparency is the "opposite of opaqueness, complexity or even secretiveness".[3]

---

[1] Bianchi (2013), p. 7.

[2] Bianchi (2013), p. 2. The author emphasizes that transparency "epitomizes the prevailing *mores* in our society and becomes a standard of (political, moral and, occasionally, legal judgement of people's conduct".

[3] Prechal and de Leeuw (2008), p. 202.

M. Vanuzzo, *Transparency in EU Law*, Studies in European Economic Law and Regulation 29, https://doi.org/10.1007/978-3-032-02345-2_2

Concepts like transparency, expressed as norms or legal principles, are unquestionably connectors between the changing social realities and the law, and they lead normative processes and interpretations of obligations.[4]

One might argue that the relevance in law of these concepts has intrinsic evidence.[5] Nevertheless, their definition and qualification are unclear.

In simple terms, transparency "means different things to different people in different contexts".[6] It implies an immediate association with information and knowledge, participatory democracy and good governance.[7]

At first glance, all these broad concepts might bring confusion when one faces the topic of the genesis and the role of transparency in law. Transparency, undeniably, is more invoked than defined.

Before addressing the issues arising from the impossibility of having a universally recognised definition of transparency in law, we could immediately describe how, and in which sense the concept of transparency in law encompasses different features and various dimensions.

In particular, transparency has been described as a broad concept that connects with the dimensions of democracy, legitimacy, quality of governance and accountability, as well as with the realisation of individual rights and market efficiencies.[8]

All these "faces"[9] contribute to an insightful view about the purposes that transparency has in law. Furthermore, they contribute to the aim of reaching a better understanding of the context in which it operates.

---

[4] See Bianchi (2013), p. 7.

[5] Buijze (2013a), p. 35. The author proposes a critical view of the intrinsic value assigned to transparency and access to information by other scholars, arguing that sometimes it builds on the grounds of their recognition as fundamental rights in authoritative legal sources or as human rights. Nevertheless, she notes that "the interpretation of a right of transparency would still depend on the extent to and the manner in which transparency contributes to the realisation of those other rights" (p. 39). Similarly, Koivisto (2016) p. 2, who argues that "the 'transparency turn' has happened without sufficient conceptual analysis or critique": "what are we calling for when we call for transparency? The methaporical authority of transparency is largely taken as self evident. At the same time, it is an ambiguous concept, which lacks a consensual definition".

[6] Bianchi (2013), p. 8.

[7] In the absence of a generally accepted definition of good governance this concept has been defined in several International Organization documents and transparency "is likewise widely accepted as an essential component of good governance". See Weiss and Steiner (2007), p. 1550.

[8] For a similar classification, see also Buijze (2013a), pp. 40–51.

[9] Buijze (2013b), pp. 3–25.

2.2 Which Purposes Does Transparency Serve Under the Theories of Transparency? Dimensions of Transparency

2.2.1 Democracy

Among the deep-seated reasons for the emergence of a need for transparency in law is the belief that transparency in decision-making processes can contribute to the democratic nature of the institutions. In this view, it is necessary to consider the strict relationship between the visibility of the public authorities' powers and the notion of democracy itself, described as the "governing of the public power in public".[10]

Transparent action taken by public authorities helps ensure that citizens are better informed about the measures of those entrusted with governmental functions, and is therefore closely linked with the democratic character of these institutions. An effective and efficient control from the citizen to the governing institutions' activities is achievable only by ensuring the publicity and the development of genuinely participatory organisational models.[11]

Therefore, one can say that the general rule concerning the democratic nature of the institutions should be transparency, and exceptions are admissible only when balancing other interests protected at a higher constitutional level.[12]

Although this overview of the transparency dimensions focuses on those that arise at the European Union level, we can briefly mention that the international law doctrine likewise recalls the democratic principle. This principle is cited in those cases where openness is the rule applicable to countries' governing bodies, contrary to the traditional diplomatic method where secrecy is the rule.[13]

---

[10] See Santini (2004), p. 4. The author quotes Norberto Bobbio's definition of democracy as "il governo del potere pubblico in pubblico" [the rule of public government in public]; see Bobbio (1980), p. 182.

[11] See Santini (2004), p. 4. The author cites the Opinion of Advocate General Tesauro delivered in the judgment C-58/94 *Netherlands v Council*, para. 14, in which he underlines that "it is scarcely necessary to point out that openness of the public authorities' action is closely linked with the *democratic nature of the institutions* inasmuch as it tends to secure better knowledge on the part of the citizens of the acts and measures—even in the phase when they are being drawn up—adopted by those who have 'government functions'. Only where there is appropriate publicity of the activities of the legislature, the executive and the public administration in general, it is possible for there to be effective, efficient supervision, inter alia at the level of public opinion, of the operations of the governing organizations and also for genuinely participatory organizational models to evolve as regards relations between the administration and the administered", emphasis added.

[12] Santini (2004), p. 5.

[13] See Santini (2004), p. 5. In the international law perspective, Bianchi (2013) p. 3 states that "the shrewdness and reputation of politicians and diplomats was measured by their ability to act discretely, striking deals behind the scenes and avoiding public scrutiny. It would be inaccurate to state those days are over. (…) But the culture all around has changed". In addition, Buijze (2013a) p. 28 refers that in the new diplomacy doctrine transparency is invoked in diplomacy to prevent the

The idea that transparency improves democracy was not implied at the beginning of the European Union legal order but received significant attention in the subsequent development of the European Union.

Since the first half of the 1990s, this theme has acquired great importance in studies on the European Union's institutional structure and decision-making processes. In particular, a significant degree of transparency of the Union's institutions and decision-making processes appears to be instrumental in overcoming the democratic *deficit* from which the Union is deemed to suffer.[14] This democratic *deficit* mainly results from the intense regulatory activity for the completion of the internal market and from the attribution of ever-growing competences to the European institutions that occurred in recent years.

As will be seen below, the greater attention to transparency led to the adoption of the Declaration No. 17 annexed to the Treaty of Maastricht and began to bear its fruits with the adoption of the Amsterdam Treaty. The Declaration No. 17 states that "transparency in the decision-making process strengthens the democratic nature of the institutions and the public's confidence in the administration".[15] It is also worth noting that the reference to transparency in decision-making processes was introduced in the Treaty of Amsterdam at the same time as the explicit mention of democracy among the principles on which the European Union is based.[16] Besides, the Treaty of Amsterdam codified access to documents as a Treaty right. Namely, Article 1 TEC stated that decisions should be taken as openly and as close as possible to citizens, and Article 255 TEC enshrined a right of access to documents held by the European Parliament, Council, and Commission.

Many "exogenous events" started these changes within the democratic dimension of transparency.[17]

The first was the Danish electorate's rejection of the Maastricht Treaty in 1992. A step forward in the following year was the adoption of a Code of Conduct by both the Commission and the Council "concerning public access to Council and

horror of another World War and that in the corporate governance doctrine transparency requires the "free flow of information between executives and stockholders". See also Frost (2003) p. 93 who quotes that "the EU has been largely the project of diplomats, and diplomats are unaccustomed to inviting the public to participate in their endeavours. Accordingly, for much of the EU's history, the public was purposefully kept at arm's length from the decision-making process".

[14] See Prechal and de Leeuw (2007), p. 51; Santini (2004), pp. 1–10; Alemanno (2014), pp. 72–74; Bestagno et al. (2024), pp. 19–22, 86–91.

[15] The Declaration adds that "the Conference accordingly recommends that the Commission submit to the Council no later than 1993 a report on measures designed to improve public access to the information available to the institutions".

[16] See Santini (2004), p. 6. To highlight the correlation between transparency and democracy, the Author also refers to the Treaty adopting a Constitution for Europe (signed in Rome on 29 October 2004), which never came into force and in which the transparency of the activities of the institutions is included in the Title dedicated to the democratic life of the Union.

[17] See Curtin (2009), p. 206.

Commission documents",[18] attempting to persuade the Danish electorate at the vote in favour of the Maastricht Treaty. Moreover, with the enlargement of the European Union in 1995 Member States such as Finland and Sweden, who had always fostered more openness, joined the EU.

Another facilitator event was the resignation of the Santer Commission that occurred in 1999 after the publication of the First Report of the Committee of Independent Expert (CIE). In that circumstance, transparency and openness and their connection to responsibility and accountability were the core of the general issue.[19]

For the purpose of the book, we shall also mention the BSE (bovine spongiform encephalopathy[20]) crisis from 1996. As will be said, the 'mad cow crisis' has shown the lack of transparency of the Committees operating within the Union. The outbreak of the BSE undoubtedly unveiled serious shortcomings in the Commission's management, with the effect of a progressive distrust in the Commission's action.[21] This event, consequently, led to an increasing awareness among citizens and regulators of the vital link between law and science and the need to reevaluate their relationship to improve the decision-making process in the EU legal order.

In this vein, the White Paper on European Governance of the 2001 mentions the food scares as an area in which there is a "perceived inability"[22] of the Union to act effectively. More specifically, the White Paper concerns "how the Union uses the powers given by its citizens" and proposes "opening up the policy-making process" to involve people and organisations in the construction of the EU policy.[23] Since this report is entirely dedicated to democracy and how people can take part in the public debate, it is of particular interest to note that in the principles underpinning good governance the first that is mentioned is *openness*, besides participation, accountability, effectiveness and coherence. In this respect, the White Paper puts forth that

---

[18] Code of Conduct concerning public access to Council and Commission documents (97/730/EC), OJ No L 340/41, 31 December 1993.

[19] See Curtin (2009), p. 206. More specifically on the resignation of the Santer Commission, Santini (2004), pp. 41 and 49.

[20] The BSE is a degenerative brain disease affecting cattle that came out for the first time in the United Kingdom in 1985. The outbreak of the BSE crisis occurred in March 1996. The temporary Enquiry Committee instituted to investigate the actions of the European and national agencies involved in the crisis presented its report in 1997. See, "Report of the Temporary Committee of Inquiry into BSE, set up by the Parliament in July 1996, on the alleged contraventions or maladministration in the implementation of Community law in relation to BSE, without prejudice to the jurisdiction of the Community and the national courts" of 7 February 1997, A4-0020/97/A, PE 220.544/fin/A in which is highlighted that "in the BSE affair there has been, at very least, a lack of transparency" and "(...) by virtue of the opaqueness, complexity and anti-democratic nature of its working, the existing system of comitology seems to be totally exempt from any supervision, thereby enabling national and/or industrial interests to infiltrate the Community decision-making process". See also Chap. 4, para. 4.1.

[21] See Vos (2000), p. 228.

[22] White Paper on European Governance, COM (2001) 428 final, OJ 27 July 2001.

[23] White Paper on European Governance, COM (2001) 428 final, 27 July 2001.

the institutions should actively communicate about "what the EU does and the decisions it takes" due to improve their "confidence".[24]

Openness and transparency are often confused and overlapped.[25] In this regard, one might propose to "unpack" the principle of openness, which has arisen as a "stand-alone principle" whose normative content may be split into the two main components: transparency and participation.[26] In this view, the latter requires the former, because participatory procedures rely on the availability of information. Moreover, an efficient participation implies the cognition of the various elements taken into account in the adopted decision.[27]

In this book, transparency and openness are concepts that could be seen as interchangeable.

We could therefore agree with those who consider these two terms as synonyms[28] and refer mainly to transparency as a critical concept to be investigated at the level of the principle itself and through the subsequent obligations laid down in EU food law.

As regards transparency as a "hallmark of good governance"[29] it is worth to mention that in the White Paper on European Governance the reference to transparency is in the chapter regarding "better policies, regulation and delivery" under the section "confidence in expert advice". Namely, the White Paper announces that "the current proposal for a European Food Authority will enhance the Union's scientific capability, *transparency* and networking in the area of food safety".[30]

In similar vein, the Code of Good Administrative Behaviour places a particular focus on good governance without defining it but identifying transparency as a requirement of good administration.[31]

---

[24] White Paper on European Governance, COM (2001) 428 final, 27 July 2001, p. 8. The European Council includes "more democracy, transparency and efficiency in the European Union" in the Laeken Declaration of the Future of the European Union.

[25] See, for the food law domain, Conte-Salinas and Wallau (2016), pp. 581–606.

[26] Alemanno (2014), p. 81.

[27] Alemanno (2014), p. 82. As a result of this "unpacking", the author refers that transparency encompasses "several subcomponents: clarity, understandability and knowledge of decisional processes, freedom of information and access to open and transparent proceedings, the duty to inform who may be seeking to influence EU decision-making, the duty to publish all legislative outputs as well as the giving of reasons" while participation "may be split into several sections: the duty to provide citizens the opportunity to make known and publicly exchange views in all areas of Union action; the duty to maintain an open, transparent and regular dialogue with the public; the duty to carry out broad consultations; as well as the duty to enable citizens to initiate and participate in ECIs".

[28] See Goudappel and Van den Brink(2011), p. 387; Heald (2006), p. 26.

[29] Curtin (2009), p. 206.

[30] "White Paper on European Governance COM (2001) 428 Final, p. 18.

[31] The European Code of Good Administrative Behavior (https://www.ombudsman.europa.eu/en/publication/en/3510) is a document drafted by the European Ombudsman that explains the meaning of the right of good administration. See also Weiss and Steiner (2007) p. 1554. According to the authors, transparency plays a role not only in the "internal dimension" of good governance—based upon good administrative behavior—but also in the "external dimension" pertaining to the

In view of the above, transparency is an important tool to *democratize* the EU, although it must be combined with measures for participation and influence over European decision-making.[32]

On the back of the strong conceptual bond between transparency and democracy it is, therefore, relevant in this book to consider transparency as an element of democracy to understand how it operates in European food law.

2.2.2 Legitimacy

The White Paper on Governance and the debate that derived from it represented a turning point in the evolution of transparency as a "means to enhance legitimacy".[33] Citizens' possibility to be informed, to understand and to take part in the decision-making process strengthens the acceptability and the legitimacy of institutions and their structures.[34] This is expressed in the pivotal *Interporc* case, where the CJEU remarks that the principle of transparency enables citizens "to participate more closely in the decision-making process, to guarantee that the administration enjoys greater *legitimacy* and is more effective and more accountable to the citizens in a democratic system and to contribute to strengthening the principle of democracy and respect of fundamental rights".[35]

Since transparency has been described as "a type of holistic medicine designed to remedy many of the ailments the body of the EU is perceived to have",[36] it has also stemmed as a means to boost the legitimacy of the European Union. It is not surprising, therefore, that the idea behind the introduction of transparency measures is a response to the legitimacy crisis of the European Union.[37]

Legitimacy stands for the acceptance of the decision-making procedures, thus it is a key feature to structure the idea that the authority is "appropriate and just"[38] in the citizens' perception. Legitimacy could be indeed defined in general terms as a

---

promotion of European values as part of the EU's foreign policy, as it is mentioned in several contexts of the Cotonou Agreement (2000/384/EC: Partnership agreement Between Members of the African, Caribbean and Pacific Group of States of the One Part, and the European Community and Its Member States on the Other Part, June 23, 2000), see p. 1558.

[32] See Frost (2003), p. 89.

[33] Alemanno (2014), p. 83.

[34] Alemanno (2014), p. 82.

[35] Judgment of 6 March 2006, *Interporc v Commission*, C-41/00 P, ECLI:EU:C:2003:125, paragraph 39. The Court refers to the principle of transparency in the form of the right of access to documents. Emphasis added. See Chap. 3, para. 3.3.2.

[36] Curtin and Meijer (2006), p. 110.

[37] See Peters (2013), p. 557. In this regard, the author recalls the Kant's quote in his Perpetual peace (1795) about transparency as a "test of legitimacy": "all action affecting the rights of other human beings are contrary to right and law if their maxim is not compatible with their being made public".

[38] De Fine Licht et al. (2014), p. 113.

"belief among the public that the appointed decision-makers have the right to make the decisions and that these decisions should be accepted".[39]

More specifically, some authors identified several meanings of legitimacy, starting from *formal* legitimacy, meaning the manner in which "a particular structure of authority was constituted and acts according to accepted legal rules and procedures".[40] As opposed to a more legal definition, *social* legitimacy alludes to the citizens' loyalty on the ground of a "strong sense of shared identity" and common interests.[41]

In this vein, it appears that transparency could be a lever to improve social legitimacy, given that the attribution of rulemaking competences to the institutions is not sufficient to create a sense of public confidence in the European Union. Hence, improving access to information and letting citizens become aware of the institution's activity may generate a sense of belonging. Openness—together with democratic values—can also lead to countering the citizens' scepticism and disaffection about the EU and its policies.[42]

In this context, we must refer to other meanings through which legitimacy can be linked to transparency. On the one hand, *output* legitimacy describes the citizens' agreement on the existence of an authority on the back of the fact that they are confident in its benefits.[43] If the results of European policies are widely published, the delivery of this information through the Internet or various channels enhances the perception of these outcomes.[44] Consequently, people may be convinced of the potential effectiveness of the European Union's activity.

On the other hand, *input* legitimacy means that citizens have a "fair chance to influence decision-making and scrutinise the results".[45] In this regard, transparency in the policymaking process allows people to see how the experts are selected and what the outcomes of their input are.[46] Nevertheless, it has been noticed that at the European Union level this is true mainly in the citizen's direct participation in the election for the European Parliament; therefore, their direct impact to input legitimacy is limited.[47]

On the political side, the main argument for transparency's ability to enhance legitimacy and public acceptance is the thought that "transparency can 'clear the fog' and quash the myths that may exist and reduce the distance between the public

---

[39] De Fine Licht et al. (2014), p. 113. The authors suggest following the Tom R. Tyler's definition of legitimacy as a "psychological property of an authority, institution or social arrangement that leads those connected to it to believe that it is appropriate, proper, and just".

[40] Curtin and Meijer (2006), p. 112.

[41] See Curtin and Meijer (2006), p. 112.

[42] See Curtin and Meijer (2006), p. 112.

[43] See Curtin and Meijer (2006), p. 117.

[44] See Curtin and Meijer (2006), p. 117.

[45] See Curtin and Meijer (2006), p. 112.

[46] See also Buijze (2013a), p. 43.

[47] See Curtin and Meijer (2006), p. 115.

and the administration at its service".[48] Similarly, the Commission launched the European Transparency Initiative (ETI) based on the consideration of high standards of transparency "as part of the legitimacy of any modern administration".[49]

With regard to the reason why openness increases legitimacy, and how, it is worthwhile spending some words about three theories of decision making: the agency theory, the deliberative democracy theory, and the procedural fairness theory.[50]

Under the agency theory, transparency "may reduce uncertainty about the agent's (i.e., the decision maker in a representative democracy) behaviour in situations characterized by moral hazard, thereby making the principal (i.e., the citizens in a representative democracy) more confident in delegating powers to the agent".[51] Accordingly, when the principal is aware of how the agents accomplishes the tasks, his trust in the agent increases.

The deliberative democracy theory, instead, asserts that a driver of legitimacy is the understanding of the reason behind decisions through public deliberation to raise citizens' willingness to accept those decisions.[52] This way may trigger legitimacy through the disclosure of the reasons and the alternative views.

The procedural fairness theory assigns to the procedure a value in itself on the back of the fact that citizens are more likely to accept decisions resulting from a fair procedure.[53] Transparency may thus lead to a clarification of the procedure that increases legitimacy.

---

[48] Communication to the Commission from the President, Ms. Walstrom, Mr. Kallas, Ms. Hubner and Ms. Fischer Boel proposing the launch of a European Transparency Initiative (European Commission (2005).The Communication follows by saying that "a high level of transparency can be instrumental in convincing the public that it is not being subjected to selective information practices ("spin"). This allows the discussion to focus on the real issues to be resolved at European level. Transparency therefore strengthens European integration by enhancing the credibility of the European institutions. Indeed, to reach its political objectives, the Commission relies both on the strength of its proposals *and* on the authority of its reputation" (p. 4).

[49] See COM (2006) 194 final 3.5.2006. The Green Paper concerning the European Transparency Initiative regards the review of the "overall approach to transparency" (see, p. 3). The aim is to "identify and stimulate a debate on areas for improvement", particularly in three "key components": the need for a more structured framework for the activities of interest representatives (lobbyists); the feedback on the Commission's minimum standards for consultation and the mandatory disclosure of information about the beneficiaries of EU funds under shared management.

[50] See De Fine Licht et al. (2014), pp. 114 ff. The Authors distinguish two types of transparency: transparency in rationale, which concerns the information on the substance of the decision and of the reasons on which it was taken; and transparency in process, which pertains to information on action "such as deliberations, negotiations, and votes that took place among and between the decision makers during the decision-making process and were thus directly fed into the decision". The latter can be a "fishbowl transparency" that occurs when the information is available in real time or a transparency in retrospect, after the decision has been made.

[51] De Fine Licht et al. (2014), p. 114.

[52] See De Fine Licht et al. (2014), p.114.

[53] See De Fine Licht et al. (2014), p. 115.

However, it is debatable whether transparency can only increase legitimacy. It has been noted that transparency can also have negative effects, for example, producing a feeling of powerlessness and frustration due to the lack of influence that people have in the decisions.[54] Moreover, there may be additional negative aspects, such as the misinterpretation of the flow of information reaching citizens and the fact that the media often prefers to showcase only the most critical aspects rather than the good news.[55]

2.2.3 *Accountability*

As shown before, the White Paper mentions accountability among all the features of good governance, such as openness. Namely, it explains the need for clarity about the roles in the legislative and executive processes, underlining the necessity that the EU institutions "take responsibility for what it does in Europe".[56]

Like transparency, accountability is a concept that is difficult to define. In its relationship with transparency, it has been argued that transparency is a *condition* for accountability; therefore, it can improve good governance.

To this extent, we can define accountability according to its "core sense" as "the process of being called to account to some authority for one's action".[57] In a similar way, accountability can be defined as "the duty of an individual or organisation to answer in some way about how they have conducted their affairs"[58], whereas

[54] See De Fine Licht et al. (2014), p. 116.

[55] See De Fine Licht et al. (2014), p. 116. The authors recalls this disappointment by referring to Otto von Bismarck' "pessimistic" quote: "Laws, like sausages, cease to inspire respect in proportion as we know how they are made". The fact that media are generally more interested in failures than in success has been claimed also by Curtin and Meijer (2006) p. 118 as a reason to question the relationship between transparency and output legitimacy.

[56] White Paper on European Governance, COM (2001) 428 final, OJ 27 July 2001, para. II. The White Paper adds that "there is also a need for greater clarity and responsibility for Member States and all those involved in developing and implementing EU policy at whatever level".

[57] Mulgan (2000), p. 555. The author asserts that "the word [*accountability*] crops up everywhere performing all manner of analytical and rhetorical tasks and carrying most of the burdens of democratic "governance". In his view, it can be identified several features of accountability: *external*, meaning that the account is given to some other person outside of the person being held accountable; it implies *social interaction and exchange* when one side seeks answers while the other side responds and accepts sanctions; and it implies also the *right of authority* over those who are accountable. He continues by analysing other less common dimension of accountability, such as "professional accountability"; "accountability as control"; "accountability as responsiveness" and "accountability as a dialogue" (see p. 556).

[58] Hood (2010), p. 989. Another effective definition of accountability is provided by Bovens (2007) p. 447, as a "relationship between an actor and a forum, in which the actor has an obligation to explain and to justify his or her conduct, the forum can pose questions and pass judgements, and the actor may face consequences".

transparency "broadly means the conduct of business in a fashion that makes decisions, rules and other information visible from outside".[59]

Apart from the duty to be accountable for one's actions, accountability also implies the duty to explain and justify one's conduct in front of a forum that judges the conduct itself and imposes sanctions.[60]

The link between these two concepts has been analysed in the doctrine as well as in the political science debate.

One of these analyses addresses this relationship in three possible ways: transparency and accountability as "Siamese twins" not distinguishable; as "matching parts" separable but complementary; and as "awkward couple", involving components in potential tension with one another.[61]

The first relationship implies that transparency and accountability are considered inextricable to the point where they cannot be distinguished as separate concepts. This happens especially in the scientific literature about good governance. However, this approach is not without criticism, since in the traditional civil system, there are several examples of how there can be accountability without transparency.[62]

On the other side, transparency and accountability can be seen as "matching parts". In this respect, the two concepts are both necessary for producing good governance. Transparency enables citizens and interest groups to be aware of the activity of the government through access to information providing "raw material" for "democratic accountability".[63] At the same time, transparency indirectly promotes government accountability by showing clearly the objectives of the governmental actions and constraining the possibility of non-noticing errors.[64]

Finally, it has been noted that there are several trade-offs between the two principles. In particular, the main argument in favour of the separation emphasises that "one-way communication" occurs when transparency acts, in place of an effective dialogue.[65] In this vein, a limitation of transparency can lead to effective governance and can prevent the risk that those who should be accountable for their actions blame avoidance rather than correctly manage their activities.

From another perspective, it has been noted that in some cases concerning access to documents the Court of Justice, on the one hand, considered transparency to be

---

[59] Hood (2010), p. 989.

[60] Similarly, Bovens (2007), p. 450. These could be also non-legal sanctions, such as political sanctions, such as the democratic elections, financial sanctions, or those affecting the reputations: see, Peters (2013), p. 567.

[61] Hood (2010), pp. 989 ff.

[62] See Hood (2010), p. 991. For instance, the author refers to the civil servants who have been answerable to their ministers without necessarily detailing their activities to the public; officeholders who can be accountable without full public disclosure about their activities; or business manager that are accountable to their superiors even without sharing all the information (e.g., market-sensitive information and commercial strategy).

[63] See Hood (2010), p. 992.

[64] See Hood (2010), p. 992.

[65] See Hood (2010), p. 992.

essential for controlling the legality of public administrations, while, on the other hand, it concluded in the opposite way by affirming that transparency damages the rule of law and accountability.[66] Several times transparency has been interpreted in a way that restricts the right of access and thereby creates a sort of accountability gap in the area.[67]

Despite efforts to analyse the link between these concepts, it should be remarked that accountability is often defined as a conceptual umbrella that also covers transparency, even if the latter is an umbrella concept itself.[68] However, the most convincing option for the aim of this book is to consider transparency as a dimension instrumental to accountability, and not a constitutive one. From this point of view, accountability is mostly used to qualify an actor's performance, so we can match the concept of responsiveness with the willingness to act transparently.[69]

Hence, it seems that the "matching-part" approach fits better for acknowledging the dimension of transparency in its link to accountability. Instruments and activities such as audits and other transparency obligations contribute to the accessibility of all of the relevant information. They also contribute to making these findings about government activities open and available.[70]

As a condition of accountability, transparency can improve the quality of governance. For instance, it allows public institutions to gain experience from past behaviour.[71] The accessibility of the information allows economic actors to select better options when operating in the market.[72] Moreover, it increases the chance that citizens will accept policies and regulations because they have understood them correctly. Again, under the dimension of accountability, transparency may serve as a tool for fighting against corruption and, in a broader sense, preserve the integrity of public officials and public institutions.[73]

In sum, transparency in relation to accountability is valuable because it helps to "alleviate the information asymmetry"[74] between the agent and the principal.

In light of the above, we could agree on an observation according to which "transparency may be considered part of accountability or a prerequisite to it: how can the public hold public authorities accountable if the public is not allowed to know what goes on within the public authorities, or if what goes on is obscure?".[75]

---

66 Adamski (2014), pp. 521 ff.

67 See Costa (2019), p. 222.

68 See Bovens (2007), p. 449.

69 In this sense, see also Bovens (2007), p. 450.

70 Buijze (2013a), p. 32.

71 Buijze (2013a) p. 46. According to Kaufmann and Bellver (2005) pp. 4 ff., the idea of transparency is closely related to accountability, since "the purpose for demanding transparency is to allow citizens, markets or governments to hold institutions accountable for their policy and performance"(p. 4).

72 Buijze (2013a), p. 46.

73 Buijze (2013a), p. 46.

74 Peters (2013), p. 567.

75 Dyrberg (2003), p. 83.

2.2.4 Realisation of Individual Rights

In order to achieve democracy, accountability and legitimacy, transparency needs to have a strong link with the individual's expectations.

In particular, transparency and the availability of information can contribute to the fulfilment of individual rights. This argument has been developed significantly in the international law perspective, where it has been argued that transparency is not only a tool intended to promote good governance by adopting information flows but also—in a rights-based perspective—a tool for promoting the rights of the people.[76] It is not new for NGOs to move precisely in this approach since they see freedom of information as a means for the enforcement of individual rights.[77]

Information rights, in this view, are considered an element of citizenship because they address the social dimension of the public in their mutual relation as well as in their relations with public authorities.[78]

The idea behind the consideration of transparency as an individual right or as a human right has largely been developed in relation to freedom of information. Namely, the international and human rights covenants grant freedom of information and declare that everyone shall have the right to *receive* information.[79] Among these, it is important to remember that Article 10 of the European Convention of Human Rights (ECHR) states that "everyone has the right to freedom of expression" and "this right shall include freedom to hold opinions and to receive and impart information and ideas without interference by public authority and regardless of frontiers (…)".

On the basis of this article, the European Court of Human Rights (ECtHR) has recognised the right of access to information assigning a broader interpretation of the concept of "freedom to receive information".[80] Moreover, in the interpretation of the Article 19 (2) International Covenant on Civil and Political Rights (CCPR)—which establishes the right to freedom of expression and information—the General Comment includes the "right of access to information held by public bodies".[81]

---

[76] See Peters (2013), pp. 586 ff.

[77] For an overview on the transparency advocacy campaigns and the "fix" that each proposes to the issues of the "state's asymmetrical information advantage over the public", see Fenster (2012), pp. 443 ff.

[78] Bovens (2002), p. 327.

[79] Peters (2013) p. 587. The author refers to: UNGA, Universal Declaration of Human Rights, A/RES/217(III)A, 10 December 1948, art. 19; UNGA, International Covenant on Civil and Political Rights, A/RES/2200(XXI), 16 December 1966 (ICCPR), art. 19 (2); European Convention for the Protection of Human Rights and Fundamental Freedom, 4 November 1950, 213 UNTS 221, art. 10; American Convention on Human Rights, 21 November 1969, 1144 UNTS 123, art. 13.

[80] See Peters (2013) p. 588. See also Österdahl (2016) pp. 68 ff.

[81] UN, Human Rights Committee, General Comment No. 34, CCPR/C/GC/34, 12 September 2011, para. 18. See Peters (2013), p. 589.

This view is also present in the Convention of the Council of Europe on access to information. In the preamble, the Convention recalls Article 10 ECHR on the freedom of expression.[82]

It seems, therefore, that the approach considering the right to be informed only in negative terms—i.e. the fact that the right prohibits governments from denying generally accessible information—has been abandoned.[83]

Transparency can then enhance individual rights, even if this is true mostly about the right of information and the consequent right of access. It is clear that transparency allows the public to make better decisions in order to achieve their personal goals. At the same time, it protects the rights of individuals against illegitimate governmental interference.[84] Both aspects seem to recall what has already been said about transparency in its link with accountability.

The idea of transparency as a means for the realisation of individual rights can also be found in the Aarhus Convention which enshrines the right of everyone to receive environmental information that is held by public authorities.[85]

Moreover, transparency can be considered a base for claims of individuals in the form of the right of access to documents of the institutions.

The right of access to documents, especially as enshrined in the specific Regulation (EC) No 1049/2001 regarding public access to European Parliament, Council and Commission document, has been defined as the milestone in the European institutions' policy on openness.[86]

The legal dimension of transparency has been studied and developed precisely in the aspect of the right of access to information. This is certainly the dimension that best captures the element of transparency that is linked to the claims of the individual. It is no coincidence, therefore, that the New Regulation (EU) No 2019/1381 on the transparency and the sustainability of the EU risk assessment in the food chain—which will be analysed further—refers to both the Regulation (EC) No 1049/2001 and the Regulation (EC) No 1367/2006 on the application of the provisions of the Aarhus Convention.

In any case, it has been argued that there is not only one type of information right, but there are several. In particular, in the relationship between citizens and public administrations, three groups of information rights can be distinguished.[87]

The first group is called "primary information rights" and concerns the rights providing citizens with immediate claims to access to authorities' information.[88] In this perspective, they might have a constitutional value, since they strengthen legal

---

[82] Council of Europe, Convention on Access to Official Documents, Trømso, 18 June 2009, Recital 2.

[83] Peters (2013), p. 588; Bovens (2002), p. 318.

[84] Buijze (2013b), p. 7.

[85] Convention on Access to information, public participation in decision-making and access to justice in environmental matter, Aarhus, 25.6.1998. See Chap. 3, para. 3.3. and Chap. 5, para. 5.2.1.

[86] See Maes (2002), pp. 199–207; Weiss and Steiner (2007), p. 1561.

[87] Bovens (2002), p. 327.

[88] Bovens (2002), p. 327.

certainty and "the process of democratic accountability".[89] Clearly, these primary information rights help to promote the democratic character and social and economic participation of a broad group of citizens.[90] The common denominator—in any case—is the right to access to public government information. The right to access information is a personal right against the administration and is defined as an administrative law right.[91]

In addition, there are the "secondary information rights" which do not grant an immediate right to access to information.[92] Nonetheless, governments allow citizens to have access to critical information channels so that these channels can convey the information to them.

Lastly, the "tertiary information rights" underpin citizens in their dealings with other individuals or other private entities, so as to capture the horizontal sphere of information.[93] In this third category, therefore, governments are in the background, and their task is to establish a framework within which citizens can exchange information.[94]

Also, this classification contributes to understand how an individual right to information has been effectively recognised, even if it is declined in several aspects.

It is precisely with respect to transparency as a right of the individual that part of the doctrine considers openness concerning the government's system as a so-called third-generation human right.[95] Accordingly, the right to access public documents should be viewed as a human right, as it can be deduced from the considerations briefly exposed above. This can be true precisely because the right of access is subsidiary to freedom of speech, so it has an "intrinsic importance", and it is confirmed by its inclusion as a fundamental right in the EU Charter of Fundamental Rights (Article 42).[96]

Given the democratic function of the right to information, this should be interpreted as broadly as possible, as well as guaranteeing access to the flow of information. In any case, this is not an unlimited right, but it is subject to legitimate restrictions, and it will be possible to restrict it on the basis of other legitimate objectives.[97]

---

[89] Bovens (2002), p. 328.

[90] Bovens (2002), p. 328.

[91] Harlow (1999), p. 286. The author divides the different kind of rights "encapsulated" in the complex concept of transparency: the "collective, democratic right to access information in which all individual citizens share" and "the personal right against the administration", at p. 286.

[92] Bovens (2002), p. 327.

[93] Bovens (2002), p. 327.

[94] Bovens (2002), p. 327.

[95] Santini, (2004), p. 10. The reference is to Debbasch (1990), p. 11.

[96] Peters (2013), p. 589. See also Stigliz (2007), pp. vii–viii. As Stigliz refers, there is a right to know about what the government is doing and why.

[97] Peters (2013), p. 590. The human right of information, in her view, is in line with the "overall rationale of human rights". Therefore, "is it the prime function of human rights to protect individuals against public power at whatever level of governance. It is a well-established principle that

It can be concluded that one of the dimensions of transparency is to guarantee individual rights that citizens can claim before other actors, both State and private. Indeed, it seems that transparency obligations—whichever way they are considered—are primarily placed in support of the interests of individuals.

2.2.5 *Market Efficiency*

Transparency can increase the level of market efficiency, as well as economic performance.[98] The flow of information facilitates decision making, because one has more elements to base its decision upon, transparency also has an impact on the management of economic resources. The availability of information influences the proper functioning of the market, and its predictability.

Information has always been seen as crucial in the economy: collecting and communicating knowledge contribute to the process of production and consumption as well as to the market organisation.[99]

As mentioned with regard to accountability and good governance, a transparent government may be more "predictable" and economic actors are therefore able to choose more wisely in the long run.[100] Thus, one of the transparency's roles might be preventing financial crises. In this view, economic policies and financial markets become more certain due to the reduced market unsureness concerning policymakers' predilections.[101] Moreover, at the international level countries with a transparent environment bear better performances in financial markets.[102]

In the context of public procurement, for example, providing the right amount of information increases competition because it extends the range of participants: "the higher the level of competition the better the market efficiency, pricing and fair market conditions".[103]

The availability of opportunities for consumers to balance commercial conditions of different undertakings and prices requires a certain amount of transparency,

---

these rights must be not only theoretical but real and effective. It is in the line with the purpose of human rights that they be opposable not only against State actor but also against inter-State ones such as international organizations, agencies and conferences", see p. 591.

[98] Buijze (2013b), p. 7. See also Chap. 5.

[99] See Mol (2006), p. 500. He distinguishes "information economy", namely the role which information plays in economic process, from "informational economy", that means "a specific form of social organization in which information generation, processing, and transmission become fundamental sources of productivity and power". On market transparency in the context of EU consumer law, see Weatherill (2013), pp. 91 ff.

[100] Buijze (2013a), p. 50.

[101] Kaufmann and Bellver (2005), p. 6.

[102] Kaufmann and Bellver (2005), p. 7.

[103] Georgieva (2017), p. 15.

which it seems to be one of the preconditions of an effective competition.[104] Thus, this enlarged competition expands the circle of participants, while also allowing the SMEs to seek and receive information on the needs of a given sector in order to actively participate in the market.[105] In the same area, it has been argued that the principle of transparency serves not only the objective of introducing openness in the Member States' public purchasing so as to prevent discrimination grounded on different nationality, but also as a "component of a system of best practice" which could add efficiency within the market.[106]

The idea of transparency as beneficial for the market and the economy in general is popular even within international organisations.[107] For instance, the WTO published an information note that recalls the purpose of transparency in the context of the COVID-19 and promotes the exchange of information between the different actors involved in supply chains.[108] This document underlines the importance of the principle clarifying that it is not only about access to information, but also confers governments and traders to be well-informed in a rapidly evolving trade dynamic.

It is clear that trade is the first to be affected by situations of uncertainty, and it is no coincidence that the reference to transparency goes precisely in the sense of emphasizing the need for it at every stage of the supply chain. In this direction, the dimension of transparency that takes as reference the market regards mainly actors other than governments. Some can classify transparency of economic markets in the WTO context as "social transparency".[109] In this view, the object pertains to the openness *outside* of the government and addresses the regulation of social activities. Namely, in the social transparency dimension, the core is not the information detained by the public actor, but the economic behaviour of citizens and companies.[110]

Since social transparency deals with public interests, it can also have a counter-interest concerning the privacy of private actors. It is admissible that the authority constrains some private action or handles them in a particular way, for instance, by granting an authorisation.[111]

In another respect, some argue that the growing focus on transparency in the food sector is linked to market issues and food security crises.[112] An analysis of how transparency is embedded within the food sector will follow. For now, it is useful to

---

[104] Georgieva (2017), p. 15.

[105] Georgieva (2017), p. 15. The author mentions that, however, there are other types of information that once available for the consumers might conduct to exceed the expected transparency level and thus limit the competition between the actors concerned.

[106] Bovis (2009), p. 289.

[107] For instance, see transparency as a trade policy action at EU institutional level: European Commission (2022) where it has been declared that the European Commission is committed to ensuring trade policy is transparent and inclusive in order to enhance legitimacy and public trust.

[108] WTO (2020).

[109] Addink (2009), p. 236.

[110] Addink (2009), p. 236.

[111] Addink (2009), p. 236.

[112] Albisinni (2015), p. 98.

point out that the market has been described as a formula that sums up three categories: innovation, globalisation and regulation.[113]

Apart from globalisation, which is a concept that has been widely considered as an internal element to both the legal order and the market, the role played by innovations and regulation has affected the economic analysis.

For instance, the regulatory framework concerning authorisations regards the innovations—such as novel food and GMOs—as having an undoubtedly strong impact on the production and the market.[114]

Moreover, in the construction of European law in general and food law in particular, the market played a decisive role. The main purpose of European rules regarding food was to encourage the free movement of goods throughout the internal market. In this regard, it is significant that the main rulings of the Court of Justice of the European Union on the internal market—among all, the pivotal *Cassis de Dijon* case, which introduced the principle of mutual recognition—concern food products.[115]

As regards the regulation, in the European law domain and in the food sector specifically, several ultra-national models of contractual relations coexist. There are not only contractual relationships from a private point of view that imply negotiations between private parties but also standards introduced by certified bodies of private origin but without an exact position in the "binding-voluntary" dichotomy.[116]

Potentially, one can look at the economic dimension of transparency as a result of a "trend" in secondary Union law.[117] In other words, it is more and more common to allude to transparency in a context not strictly connected with authority's information on security threats.[118] Hence, transparency seems to go beyond the unilateral exchange of information—from the authority to the citizens—in relation to risk. It is a concept increasingly suitable for different fields and activities. One can also look at it as a positive value to dimensions such as the normal dynamics of the market and the contractual relationships.

2.3 Uncertainties on the Definition of Transparency

As can be seen, transparency is a broad notion connected to different concepts such as democracy, legitimacy and accountability. It is a means to satisfy individual rights, and it can improve the market's efficiency. Since transparency has a link with

---

[113] Albisinni (2015), p. 100.

[114] On the link between transparency, innovation and authorisations, see Chap. 5.

[115] Judgment of the Court of 20 February 1979, *Rewe-Zentral AG v Bundesmonopolverwaltung für Branntweinudgment*, C-120/78, ECLI:EU:C:1979:42.

[116] Albisinni (2015), p. 100.

[117] Conte-Salinas and Wallau (2016), p. 582.

[118] Conte-Salinas and Wallau (2016), p. 582.

the dimensions discussed above, it seems almost impossible to find a definition suitable for all the purposes transparency has.

Openness leads to values and principles that have a broad consensus among citizens. Nonetheless, this does not indicate that citizens have a clear perception of what precisely transparency means and how to grasp it.[119]

It is popular among scholars to associate transparency with the metaphorical image of a window or a glass through which one can look.[120] In particular, transparency has been defined as "the ability to look through the windows of an institution",[121] or "lifting the veil of secrecy".[122] This means, for instance, that information or processes need to be "open and available for examination and scrutiny" in order to be transparent.[123]

In the policy debate, it is common to associate transparency with governmental openness, such as an institutional *relation* between an actor and the public.[124] In this vein, the term arises in the obligation of public authoristies to be transparent. Public authorities can be bearers of transparency about their own affairs, or they can create transparency, for instance, by making data available.[125]

In the legal doctrine as well as in the European institutions' documents, one can observe several positions on the issue.

In a restrictive conception, the problem of transparency is identifiable with the question of public access to the institutions' documents.[126] As already mentioned, and as will be seen later, this is only one of the aspects—although the most developed among legal scholars—in which transparency is expressed.

---

[119] Santini (2004), p. 8. The author recalls Curtin (1996), p. 95 ff. who states that the absence of an extensive debate "on the manner in which more openness may genuinely contribute to a reduction of the democratic deficit in the EU system of governance".

[120] Curtin (2009), p. 204. According to the author, "the term transparency evokes the image of a clear pane of glass through which the sunshine (or the light) can beam in an unrestrained fashion". See also Bianchi (2013) p. 9: "(…) the most effective metaphor to convey its fundamental idea is that of a clean window that you can look through". The Author questions the elusiveness of this idea, suggesting some critical questions: "Do I have a clear view if I look through a clean window? What if my view is blurred? Is what I see reality or just a mental representation of what I expect to see? (…) Is what I see not dependent also on the position in which I am behind the window?" (see p. 9). In addition, the author concludes that "despite its fallacies, we still find the image of the window a good example of what transparency stands for" but it can also be "misleading" or turn our perception into an "illusion" (see p. 10 and 15).

[121] Den Boer (1998), p. 105. See also Meijer (2013), p. 430.

[122] Curtin (2009), p. 204. According to Mock (2000), p. 294, in the legal theory transparency is a term of "analogy". If the law and its commands can be understood by the person subjected to it so as to comply, then the law is transparent and not "opaque".

[123] Schauer (2011), p. 1343. The author recalls the Oxford English Dictionary which defines transparency as "the property of transmitting light, so as to render bodies lying beyond completely visible".

[124] Meijer (2013), p. 430; Mock (2000), p. 300.

[125] Buijze (2013a), p. 30.

[126] See Santini (2004), p. 8.

According to a different doctrine, the concept of transparency means a general citizens' need of comprehensibility of the structure and decision-making mechanisms of the European Union.[127] In this vein, transparency refers not only to access to executive-held information but also—more widely—to an "open government" as such.[128]

In between these two positions, there are the most varied conceptions and understandings of the notion. These mostly focus on the procedures considered from time to time more functional to ensure adequate flows of information on the activities and decision-making processes of the European Union.[129]

Even outside the context of the European integration process, this variety of opinions has led to three main conclusions: those who deny that transparency can be defined from a legal standpoint, those who give it a legal meaning instead, and those who assign to transparency a pragmatic denotation as a way to measure the performance of authorities' actions, although not included in a precise legal institution.[130]

The question of whether transparency can be considered as a principle in the European Union's legal system will be addressed later in Chap. 3.

Since the concept of transparency is flexible and "polysemic",[131] we consider the simple and concrete meaning which appears to be more appropriate in this context: *the availability of information* as well as *the exchange of information between different actors.*[132]

Information is an important element of transparency and can be defined as "the processes data which bears a reasonable possibility of altering the world perception of someone receiving the data".[133] In this way, information is a key concept, and it

[127] Santini (2004), p. 8.; the reform proposal undergone to the Intergovernmental Conference for the revision of the Treaty of Maastricht invoked this denotation of transparency.

[128] Curtin (2009), p. 205.

[129] For instance, Santini (2004), p. 9, recalls the Italian administrative doctrine: one part recognises the value of transparency as a metaphorical word, devoid of any legal technical meaning. In this sense, some authors make the meaning of transparency coincide with that of advertising. On the contrary, others consider transparency as a general principle of constitutional value in the Italian legal system. Moreover, there are also those who place themselves in an intermediate position: they recognise the legal value of transparency even if it is not "crystallized" in a particular legal instrument.

[130] Santini (2004), pp. 9 ff.; Albisinni (2015), p. 97; Lequesne (1998), p. 13.

[131] Albisinni (2015), p. 97. According to Mendes (2020), p. 7, the "ambivalence" and vagueness of the concept in combination with the institutional or academic discourses built around the role of transparency make it difficult to identify the reasons why in the process of EU integration transparency "was either an important concern or at the forefront of policy making".

[132] For the concept of availability of information as an element of the transparency's notion, see Mock (1999) p. 1080; Santini (2004), p. 9; Schauer (2011), p. 1343; Georgieva (2017), p. 17; Meijer (2013), p. 430; "if transparency stands for availability of information, by all means the EU is a transparent polity" (Mendes (2020), p. 2).

[133] Mock (1999), p. 1075. For Mock, there are mainly four elements of transparency: the governmental action; the availability of information; the audience that receive the information and the accuracy and clarity as information's features (see at p. 1079 ff.).

is possible to look at it as an object of *interaction* between different subjects and not as a static element.

In the literature, the availability of information included in the concept of transparency can encompass not only the duty to make information accessible to those who seek it but also a more general demand for "visibility" of information.[134] Using the aforementioned metaphorical image of glass, we expect transparency to "open a window" on information.[135]

In this vein, transparency has been defined as "the state that occurs if people can easily ascertain and understand the state of the world and predict how their own actions will affect that world".[136] Accordingly, transparency is a "quality" rather than a measure—or, specifically, the "measure of the degree to which information about official activity is made available to an interested party".[137] The latter definition takes indeed into account the "official" activity from public authorities. This view is narrower than what transparency as a comprehensive legal tool seems to require and it can lead only to a meaning of transparency that has been successful in the political debate, as mentioned above. However, this definition helps in understanding the dynamics of the concept.

A similar view that builds on the availability of information and puts it in the context of the international legal system defines transparency as "a concept (…) to ensure the *public's right to the availability and accessibility of a certain level of information* about (institutional) norms, rules, procedures and regimes, and the actions of participants; where the information provided should be presented in an understandable and clear manner and should always be sufficient to facilitate monitoring, verification and assessment".[138]

It is worth mentioning this definition as it adds the aspect of the comprehensibility of the information, that is, information must be understandable by the recipients. It further takes into account the effects of transparency on the actions of those who receive the information. In this regard, some argue that transparency must embody

---

[134] See Mock (1999), p. 1080; Curtin (2009), p. 205. For instance, Bianchi (2013) p. 8 quotes the definition of transparency given by the United Nations Economic and Social Commission for Asia and the Pacific (ESCAP) (2009) because of its focus on the information: "Transparency means (…) that information is freely available and directly accessible to those who will be affected by such decisions and their enforcement. It also means that enough information is provided and that it is provided in easily understandable forms and media".

[135] Bianchi (2013), p. 10.

[136] Buijze (2013a), p. 31.

[137] Mock (1999), p. 1082. The author has given another—broader—definition of transparency in the legal context as s "a measure of the degree to which the existence, content, or meaning of a law, regulation, action, process, or condition is ascertainable or understandable by a party with reason to be interested in that law, regulation, action, processes or condition". For a comment on the definitions given by Mock, see Buijze (2013a) pp. 30 ff.

[138] Georgieva (2017), p. 13. Emphasis added.

not only the "visibility" of the information, but also its "inferability" so the information can lend itself to "verifiable inference or conclusions".[139]

Nevertheless, it is not effective for the purpose of this study to go further into the various definitions offered by the doctrine. These vary depending on peculiar approaches and contexts from clear procedures through the decision-making process to full access and so on.[140] Instead, the emphasis on the release of the information builds on the importance of effective communication among the parties involved. Furthermore, it should be considered not only the availability of information but also the *exchange* of this information.

In the legal context and, in particular, in the specific field of food law, it will not be taken into account the only—vertical—perspective of transparency from public authority towards citizens. To grasp how the *availability* and the *exchange of information* shape the concept of transparency within this book, transparency must be seen as a tool serving different purposes.

In particular, the definition of transparency should be connected with the scope of investigating whether a principle of transparency in European food law exists and how the related obligations are incumbent on public authorities and private actors. In doing this, we must agree with the authors that consider transparency as an *instrument for a purpose*[141] and even accept it as an attribute or a quality "with uncertain content".[142]

If transparency relates to information flows between subjects, it is possible to describe this dynamic in various ways.

Some authors propose a classification among the uses of transparency, distinguishing between the hypotheses in which it is emphasised in *horizontal relationships* of a contractual nature and the cases in which transparency is articulated in *vertical power* relationships.[143] Within the latter category, it is then possible to separate the transparency of the governed towards the governors and its opposite—the transparency of the governors towards the governed.[144]

Given this dynamic, it is a fact that the possible relationships and interactions within European law in general and food law specifically regard different actors. The Farm to Fork strategy is a clear example. The strategy emphasises that European Green Deal's aim to transition to a more sustainable food system. This purpose

[139] Michener and Bersch (2013), p. 234, according to whom "the demand of inferable information not only stems from the recognition that 'raw' data often permits greater verifiability and is more modular, but also that false transparency and unintelligible disclosures remain enduring problems" (at p. 236).

[140] Guina-Dorman (2007), p. 280.

[141] Bianchi (2013), p. 18; Peters (2013), p. 558.

[142] Prechal and de Leeuw (2007), p. 61.

[143] Rideau (1998), pp. 1 ff. See Santini (2004), p. 12. In a similar way, see Heald (2006), pp. 25–43 and Buijze (2013a), p. 33.

[144] Rideau (1998) pp. 1 ff., Santini (2004), p. 12.

requires a "collective approach" involving public authorities, private-sector actors, non-governmental organisations, social partners, academics and citizens.[145]

We can already advance that in the context of food law, for example, one might reasonably question whether the administrative transparency principle underpinning the Regulation (EC) No 1049/2001 on access to documents is of the same nature as the principle regulating risk communication to consumers as provided by Article 6 of the Regulation (EC) No 178/2002 (General Food Law).[146]

Given this context and after having analysed the above-mentioned doctrinal definitions, it seems appropriate to define transparency plainly as the *availability and exchangeability of information between different actors in a specific context, where the information provided is understandable, and the recipients can direct their actions consequently.*

Even if this broad and potentially vague definition may invite to criticism, it seems the most suitable for capturing the flexibility of the notion in the context of food law. Nevertheless, it is reasonable to state that there may be several notions of transparency that can fit in the different contexts taken into consideration. One of the most cited definitions of transparency for the food sector relates to the supply chain network. Namely, transparency for a supply chain network is "the extent to which all the network's stakeholders have a shared *understanding* of, and *access* to, product and process related information that they request, without loss, noise, delay and distortion".[147] Even without going into detail, this definition illustrates that transparency requires information to be available—therefore, accessible—and understandable for all the actors interested.

Despite the absence of a shared definition of transparency, given the adaptability of the notion to multiple contexts, this book explores whether there is a general principle of transparency in European Union law and in food law particularly. To this end, the analysis turns to how the principle operates in practice, focusing on some specific transparency obligations that are incumbent on the actors from time to time considered. As regards the broader European Union law context, transparency in a vertical sense, i.e. in the relationship between public authorities and citizens, offers a valuable angle. As far as food law is concerned, the focus shifts to how transparency duties can directly concern other actors, such as food business operators.

In terms of the meaning of transparency obligation, it seems effective to recall the arguments proposed by a scholar who points out that a transparency obligation can be analysed by answering a set of questions, such as: "what is the reason for the

---

[145] Communication from the Commission to the European Parliament, the Council, the European Economic and Social Committee and the Committee of the Regions, A Farm to Fork Strategy for a fair, healthy and environmentally-friendly food system, COM/2020/381 final, 20.5.2020. On the farm to fork strategy, see Schebesta and Candel (2020), pp. 586–588.

[146] Regulation (EC) No 178/2002 of the European Parliament and of the Council of 28 January 2002 laying down the general principles and requirements of food law, establishing the European Food Safety Authority and laying down procedures in matters of food safety, OJ L 31, 1.2.2002.

[147] Beulens et al. (2005), p. 482. Emphasis added.

obligation? Who must be transparent? To whom? About what? When? Actively or passively? Are there quality standards for the information that is being supplied? What are the exceptions?"[148]

All these questions are useful both to understand the notion of transparency and to identify the transparency instruments in the legislation examined.

2.4 Target of Transparency

As mentioned above, the issue of transparency arose and entered into law in the form of a tool that can enhance other values, such as democracy, legitimacy, and accountability. Moreover, it can enhance individual rights and improve market efficiency.

As regards transparency obligations, we must consider the subject bearing these obligations. That is the "target" of transparency.

Since the issue of transparency has had its origin and has been more extensively studied in the relations between citizens and governments, transparency is regarded as a "classic" tool for monitoring the compliance of the states or public authorities in general.[149] To this extent, we should briefly recall that the target of transparency is typically the public authority because of its link with the democracy dimension.

In particular, it has been argued that transparency is relevant in a democracy because the relationship between citizens and institutions is grounded on "reasons of control and trust".[150] Namely, according to the control-based approach, transparency obligations are delegated to the public authority because only the consent of the people can legitimate the control that the authority pursues to exercise.[151] The mechanism of control aims to ensure the realisation of citizens' interests and intends to limit the discretion in the public authority's activities.[152] In this vein, the availability of information becomes a means of "making visible" whether the government activity corresponds to the citizens' interests: transparency is then necessary to enable people to formulate opinions and conditions for consent.[153] Therefore, the target of openness is primarily public authorities because they must be subjected to citizens' control to meet the democratic requirement.

Another—complementary—approach grounds the authority's legitimacy in the citizens' trust. This argument is popular among the doctrine that sees in transparency a response of the European institutions to the democratic deficit.[154] Accordingly,

---

148 Buijze (2013a), p. 35.

149 Peters (2013), p. 549.

150 Abazi and Tauschinsky (2015), p. 78 ff.

151 Abazi and Tauschinsky (2015), p. 79.

152 Abazi and Tauschinsky (2015), p. 80.

153 Abazi and Tauschinsky (2015), p. 84.

154 See para. 2.2.1.

putting trust as a base of the conceptual relation between governors and governed aims to create a system of "cooperative relations" between European states and citizens.[155] In this vein, even the role of the information can be described in a different way. Information is indeed necessary to sustain the relation of trust in which "deference and authority" are premised because trustworthiness "connotes the degree to which a person can be trusted".[156] Therefore, in this approach information has the different function of establishing the conditions upon which the two-way relationship between authorities and citizens are supported.[157]

Even from this point of view, one can conclude that trust is the basis of an approach that considers the authority as the primary recipient of the rules of transparency.

In any case, whether one wants to base the legitimacy and role of information on the control relationship or whether one intends to establish it on trust, this does not change the idea behind the origin of transparency. Especially in the context of the process of European integration, and in the consequent issue of the democratic deficit derived from it, transparency obligations point to the public authority.

Despite this, it can be argued that transparency is increasingly "multi-directional".[158] This indicates that recently not only citizens require that governments release information on their activity, but also authorities have begun to demand greater transparency from non-profit organisations as well as private sector actors.[159] Hence, the availability of information is "demanded *from* but also provided *to*"[160] private actors and citizens. As mentioned with regard to the economic dimension of transparency and the market efficiency argument,[161] governments can also enhance transparency through a "translucency practice" in society, especially in the economic behaviour.[162] Governments, thus, impose the transparency principle to citizens and companies in their activities and in the regulations of such activities.

Another point of view that seems to cast doubt on whether transparency obligations are addressed only to the authorities comes from international law. In this field, it has been argued that non-governmental organisation (NGOs) and commercial enterprises have stepped in as "new transparency powerbroker".[163]

As regards NGOs, they have been criticized for being "undemocratic" or for not reaching an adequate level of independence from the governments,[164] hence the

[155] Abazi and Tauschinsky (2015), p. 81.

[156] Abazi and Tauschinsky (2015), p. 84.

[157] Abazi and Tauschinsky (2015), p. 85.

[158] Michener and Bersch (2013), p. 235.

[159] Michener and Bersch (2013), p. 236.

[160] Gupta et al. (2020), p. 84.

[161] See para 2.2.5.

[162] Addink (2009), p. 236.

[163] Peters (2013), p. 551.

[164] Peters (2013), p. 551.

push to become more transparent. To solve this legitimacy issue, they must comply with transparency obligations.

Another point of discussion involves *private or economic actors*. In their respect, the question on whether they can be legitimately recipients of transparency obligations creates some concerns. For instance, they have the right to privacy and business secrecy.[165] Contrary to the states, whose main task is to pursue the public interest, economic actors adhere to activities in their own interest with the goal of generating welfare by supporting employment and providing goods.[166] However, these considerations do not seem sufficient to overlook how individuals may be subject to transparency obligations. Indeed, emphasising the commonalities and differences between public authorities and private actors as potential recipients of transparency rules does not seem to align with the aim of the book.

Rather, a central figure in food law is food business operators, and in particular those of a private nature. Pursuant to Article 3(3) of the Regulation (EC) No 178/2002 (the General Food Law[167]), food business operator (FBO) means "the natural or legal person responsible for ensuring that the requirements of food law are met within the food business under their control". Therefore, in the food law domain, it should be considered another kind of relation: the one between consumers and food business operators. Using the classification mentioned above regarding transparency in horizontal or vertical relations, the interaction between consumers and food businesses falls under the former.

On the other side, it is clear that the bond between authorities and citizens results in a "vertical power" relationship. According to what is expressed above, this relation is grounded on the two elements of "control" and "trust". However, trust is a factor that occurs also in the relations between consumers and food businesses.[168] Given the increasing attention of consumer trust in the market relations and in the food law in general, a relevant consequential question is whether and to what extent specific private actors—such as food business operators—are required to comply with transparency obligations.

References

Abazi V, Tauschinsky E (2015) Reasons of control and trust: grounding the public need for transparency in the European Union. Utrecht Law Rev 11:78–90. https://doi.org/10.18352/ulr.319

Adamski D (2014) Access to Documents, Accountability and the Rule of Law—Do Private Watchdogs Matter? Eur Law J 20:520–543

---

[165] Peters (2013), p. 551.

[166] Peters (2013), p. 551.

[167] Regulation (EC) No 178/2002 of the European Parliament and of the Council of 28 January 2002 laying down the general principles and requirements of food law, establishing the European Food Safety Authority and laying down procedures in matters of food safety, OJ L 31, 1.2.2002.

[168] See de Boer (2019), p. 253.

Addink GH (2009) The transparency principle in the framework of the WTO. Indonesian J Int Law 6(2):232–243

Albisinni F (2015) Transparency, Crisis and Innovation in EU food law, Rivista Della Regolazione Dei Mercati 1:97–115

Alemanno A (2014) Unpacking the principle of openness in EU law. Transparency, participation and democracy. Eur Law Rev 39(1):72–90

Bestagno F, Draetta U, Santini A (2024) Elementi di Diritto dell'Unione Europea. Profili istituzionali. Giuffrè Francis Lefebvre, Milano

Beulens AJM, Broens D-F, Folstar P, Hofstede GJ (2005) Food safety and transparency in food chains and networks Relationships and challenges. Food Control 16(6):481–486

Bianchi A (2013) On Power and Illusion: the Concept of Transparency in International Law. In: Bianchi A, Peters A (eds) Transparency in International Law. Cambridge University Press, Cambridge

Bobbio N (1980) La democrazia e il potere invisibile. Rivista Italiana di Scienza Politica 10:181–203

Bovens M (2002) Information rights: citizenship in the information society. J Polit Philos 10:317–341

Bovens M (2007) Analysing and assessing accountability: a conceptual framework. Eur Law J 13:447–468

Bovis CH (2009) The effects of the principles of transparency and accountability on public procurement regulation. In: Hofmann H, Türk AH (eds) Legal challenges in EU administrative law. Towards an integrated administration. Edward Elgar Publishing, Cheltenham, pp 288–321

Buijze A (2013a) The principle of transparency in EU law. Dissertation, Utrecht University

Buijze A (2013b) The six faces of transparency. Utrecht Law Rev 9(3):3–25

Conte-Salinas N, Wallau R (2016) The concepts of transparency and openness in European food law. In: Steier G, Patel KK (eds) International food law and policy. Springer International Publishing, Cham, pp 581–606

Costa M (2020) Accountability Through Transparency and the Role of the Court of Justice of the European Union. In: Hirsch Ballin E., van der Schyff G., Stremler LLM, M. (eds.), European Yearbook of Constitutional Law 2019 221–242

Curtin D (1996) Betwixt and between: democracy and transparency in the governance of the European Union. In: Winter JA, Curtin D, Kellermann AE, de Witte B (eds) Reforming the treaty on European Union. The legal debate. Kluwer Law International, The Hague/Boston/London, pp 95–130

Curtin D (2009) Executive power of the European Union: law, practices, and the living constitution. Oxford University Press, Oxford, pp 204–206

Curtin D, Meijer A (2006) Does transparency Strenghten legitimacy? A critical analysis of European Union policy documents. Inf Polity 11:109–122

De Boer A (2019) Transparency and consumer Trust in Scientific Assessments under European Food Law. In: Urazbaeva A, Szajkowska A, Wernaart B, Tilkin Franssens N, Spirovska Vaskoska R (eds) The functional field of food law. Reconciling the market and human rights. European Institute for Food Law Series, Wageningen Academic Publishers, Wageningen, pp 245–255

De Fine Licht J, Naurin D, Esaiasson P, Gilljam M (2014) When does transparency generate legitimacy? Experimenting on a context-bound relationship. Governance 27:111–134

Debbasch C (1990) Introduction. In: Debbasch C (ed) La transparence administrative en Europe. Centre National de la Recherche Scientifique, Paris

Den Boer MGW (1998) Steamy windows. Transparency and openness in justice and home affairs. In: Deckmyn V, Thomson I (eds) Openness and transparency in the European Union. European Institute of Public Administration, Maastricht, pp 91–105

Dyrberg P (2003) Accountability and legitimacy: what is the contribution of transparency? In: Arnull A, Wincott D (eds) Accountability and legitimacy in the European Union, 1st edn. Oxford University Press, Oxford, pp 81–96

European Commission (2005) Communication to the Commission from the President, Ms Wallström, Mr Kallas, Ms Hübner and Ms Fischer Boel proposing the launch of a European transparency initiative. Available at https://www.statewatch.org/media/documents/news/2005/nov/com-transparency-nov-05.pdf. Accessed 7 June 2025

European Commission (2022) Factsheet on Transparency Policy in DG Trade Available at https://circabc.europa.eu/ui/group/7fc51410-46a1-4871-8979-20cce8df0896/library/ebae888d-6243-48a2-a766-e21923793a71/details. Accessed 7 June 2025

Fenster M (2012) The transparency fix: advocating legal rights and their alternatives in the pursuit of a visible state. Univ Pittsburgh Law Rev 73:443–503

Frost A (2003) Restoring faith in government: transparency reform in the United States and the European Union. Eur Public Law 9:87–104

Georgieva I (2017) Using transparency against corruption in public procurement: a comparative analysis of the transparency rules and their failure to combat corruption. Springer, Cham

Goudappel F, Van den Brink A (2011) Processes of transnationalization of administrative values: administrative regulation and transparency in the EU. In: Anthony G, Auby JB, Morison J, Zwart T (eds) Values in global administrative law. Hart Publishing, Oxford and Portland, pp 373–392

Guina-Dorman D (2007) Disclosure of information on food control inspections. Eur Food Feed Law Rev 2:279–289

Gupta A, Boas I, Oosterveer P (2020) Transparency in global sustainability governance: to what effect? J Environ Policy Plan 22:84–97

Harlow C (1999) Freedom of information and transparency as administrative and constitutional rights. Cambridge Yearb Eur Legal Stud 2:285–302

Heald D (2006) Varieties of transparency. In: Hood C, Heald D (eds) Transparency, the key to better governance? Oxford University Press, Oxford, pp 24–43

Hood C (2010) Accountability and transparency: Siamese twins, matching parts, awkward couple? West Eur Polit 33:989–1009

Kaufmann D, Bellver A (2005) Transparenting transparency: initial empirics and policy applications. SSRN Electronic Journal available via. https://doi.org/10.2139/ssrn.808664. Accessed 24 Dec 2024

Koivisto I (2016), The anatomy of transparency: the concept and its multifarious implications. EUI MWP 9: Available via https://hdl.handle.net/1814/41166

Lequesne C (1998) La Transparence: Vice ou Vertu Des Democraties? In: Rideau J (ed) La Transparence Dans l'Union Europénne: Mythe Ou Principe Juridique? L.G.D.J, Paris, pp 11–18

Maes M (2002) The "New" regulation on access to documents. In: Deckmyn V (ed) Increasing transparency in the European Union? European Institute of Public Administration, Maastricht, pp 199–207

Meijer A (2013) Understanding the complex dynamics of transparency. Public Adm Rev 73:429–439

Mendes J (2020) The principle of transparency and access to documents in the EU: for what, for whom and of what? Universitè Du Luxembourg, Faculty of Law, Economics and Finance Law Working Paper Series, no. Paper number 2020-004. Available via https://www.ssrn.com/abstract=3557795

Michener G, Bersch K (2013) Identifying transparency. Inf Polity 18:233–242

Mock WBT (1999) On the centrality of information law: a rational choice discussion of information law and transparency. J Comput Inf Law 17(4):1069–1100

Mock WBT (2000) An interdisciplinary introduction to legal transparency: a tool for rational development. Dickinson J Int Law 18:293–304

Mol APJ (2006) Environmental governance in the information age: the emergence of informational governance. Environ Plan C Gov Policy 24:497–514

Mulgan R (2000) 'Accountability': an ever-expanding concept? Public Adm 78(3):555–573

Österdahl I (2016) Transparency as part of a European rule of law. In: Schroeder W (ed) Strengthening the rule of law in Europe: from a common concept to mechanisms of implementation. Hart Publishing Ltd, Oxford, pp 61–79

Peters A (2013) Towards Transparency as a Global Norm. In: Bianchi A, Peters A (eds) Transparency in International Law. Cambridge University Press, Cambridge, pp 534–607

Prechal S, de Leeuw M (2007) Dimensions of transparency: the building blocks for a new legal principle? Rev Eur Adm Law 0(1):51–62

Prechal S, de Leeuw M (2008) Transparency: A General Principle of EU Law? In: Bernitz U, Nergelius J, Cardner C (eds) General principles of EC Law in a Process of Development. Kluwer Law International, Alphen aan den Rijn, Zuid-Holland, pp 201–239

Rideau J (1998) Jeux d'ombres et de Lumiers En Europe. In: Rideau J (ed) La Transparence Dans l'Union Europĕnne: Mythe Ou Principe Juridique? L.G.D.J, Paris, pp 1–9

Santini A (2004) Il principio di trasparenza nell'ordinamento dell'Unione Europea. Giuffrè, Milano

Schauer F (2011) Transparency in three dimensions. Univ Ill Law Rev 4:1339–1358

Schebesta H, Candel JJL (2020) Game-changing potential of the EU'S farm to fork strategy. Nat Food 1:586–588

Stigliz JE (2007) Foreward. In: Florini A (ed) The right to know: transparency for an open world. Columbia University Press, New York, pp vii–viii

United Nation Economic and Social Commission for Asia and the Pacific (ESCAP) (2009) What is good governance? Available at https://www.unescap.org/resources/what-good-governance. Accessed 7 June 2025

Vos E (2000) EU food safety regulation in the aftermath of the BSE crisis. J Consum Policy 23:227–255

Weatherill S (2013) EU consumer law and policy. Elgar European Law Series, Cheltenham

Weiss F, Steiner S (2007) Transparency as an element of good governance in the practice of the EU and the WTO: overview and comparison. Fordham Int Law J 30:1545–1586

WTO (2020) Transparency – Why it matters at times of crisis. Available at www.wto.org/english/tratop_e/covid19_e/transparency_report_e.pdf. Accessed 18 Apr 2025

Chapter 3
Is There a General Principle of Transparency in EU Law?

Abstract This chapter describes the EU law doctrinal approach on transparency and the most relevant CJEU case law on the subject matter. The scope is to answer the sub-question concerning the existence of a general principle of transparency in EU law. Before addressing the topic of access to documents, the chapter traces the development of the principle of transparency in different fields of EU law and the difference between active and passive transparency. Furthermore, in EU law transparency also emerged next to other well-established legal principles, such as sound administration and legal certainty. Regulation (EC) No 1049/2001 on access to documents and Regulation (EC) No 1367/2006 seem to remain the core legal framework of transparency in EU law, as exposed at the end of this chapter.

3.1 Genesis and Evolution of the EU Legal Principle of Transparency

As seen with regard to the dimension of democracy, transparency in the European Union has received great attention starting from the 1990s. In the wake of the Treaty of Maastricht the notion of transparency was "prayed in aid" to overcome the democratic *deficit* and the citizens' failure to "bond" with European institutions.[1]

As observed, transparency was a concern before the Treaty of Maastricht, but it explicitly became one of the themes regarding European policies only after it. It has been argued that the principle of transparency for decades was not an issue: a person could have access to certain documents held by the institutions only where his or her personal interests were affected by a specific procedure.[2] For instance, in a pronouncement of the Court of Justice of 1991 (*BEUC v. Commission*), the Court supported the Commission's refusal to grant access to documents to an organisation

[1] Harlow (1999), p. 288.

[2] Lenaerts (2004), p. 318.

M. Vanuzzo, *Transparency in EU Law*, Studies in European Economic Law and Regulation 29, https://doi.org/10.1007/978-3-032-02345-2_3

defending consumers' interests on the ground that the proceedings thereof were not directed against practices attributable to consumers defending their interests.[3]

From the birth of European communities to the beginning of the 1990s, transparency—intended as the possibility by private individuals to know the activity and decision-making processes of the institutions—regarded norms and principles directly pursuing *other aims*.[4] The duty to give reasons, the right to access to the file as an expression of the right of defence and the political controls made by the European Parliament can be included in these aspects of transparency in EU law.[5]

The growing citizens' distrust in governments and a renewed attention on governments' activity by civil society and non-governmental organisation spurred the process of increasing transparency in decision-making at the European Union level.[6] Within this context, the Declaration n. 17 to the Treaty of Maastricht on access to information underlined that transparency in decision-making process strengthens the democratic nature of the institutions.[7] Accordingly, the Conference issuing the Declaration recommended that the Commission submit to the Council a report of measures set up to improve public access to the information available to the institutions no later than 1993.[8] The outcomes of this declaration led to a series of position papers and consultations issued by the Commission. It emerged that the chosen way forward was the promulgation of a code of conduct on access to documents.[9] On 6 December 1993, the Council and the Commission approved a Code of Conduct establishing the principle of "the *widest possible access* of the public to documents held by the Commission and the Council".[10]

The provisions of the Code were later included in Council Decision 93/731 and in Commission Decision 94/90 regarding public access to Council and Commission

---

[3] Judgment of the Court of 28 November 1991, *Bureau européen des unions des consommateurs (BEUC) v Commission of the European Communities*,C-170/89, ECLI:EU:C:1991:450; see Lenaerts (2004), p. 318.

[4] Santini (2004), p. 271.

[5] See Santini (2004), p. 271 ff. According to the author, the publicity and the notifications of the legal acts, the duty to give reason and the Parliament's controlling powers are part of a body of rules and principles that exist even before the Maastricht Treaty and which aim to 'implicitly' achieve transparency. For an in-depth analysis of these institutions, see pp. 20–73. See also Vesterdorf (1998), pp. 903 ff.

[6] Lenaerts (2004), p. 318.

[7] See Chap. 2, para. 2.2.1.

[8] Tomkins (1999), p. 222.

[9] The proposals for such a code were set out in the Commission paper "Openness in the Community" and then discussed by the European Council at its meeting in Copenhagen. See Tomkins (1999), p. 222.

[10] Code of conduct concerning public access to Council and Commission documents, 93/730/EC. Official Journal L 340/41 31/12/1993. The Preamble states that "Considering is desirable to establish by common agreement the principles which will govern access to Commission and Council documents, it being understood that it is for each of them to implement these principles by means of specific regulations", therefore the Code is not intended to produce legal effects. See Santini (2004), p. 188.

documents. With respect to the documents of the European Parliament, the Decision 97/632 implemented transparency in the form of access.[11] Notwithstanding that different decisions were directed to give effect to the principle of the broad possible access to documents, they contained some exceptions to the rule of access to protect public or private interests, such as public security and monetary stability and protection of business secrets.[12]

It has been argued that the decisions of the Council, the Commission and the European Parliament on access to documents were not adopted on the basis of a general principle of European law as such. The three institutions based these decisions on the back of their power of internal organisation.[13] However, in proceedings in front of the European Courts the question of whether the principle of transparency constituted a general principle of EU law soon emerged.[14]

Recognising a principle of transparency would empower the Courts not only to verify if a refusal to grant access to a specific document is based on the particular exceptions included in the decision of the institution concerned, but also to "assess the legality" of the different exception themselves in the light of a general principle of transparency.[15] That's why the question has been of particular importance since the adoption of the Code of Conduct.

Before recalling what scholars have specified as regards the existence of a principle of transparency in EU law, it is necessary to make a distinction between "active" and "passive" transparency.[16]

---

[11] Lenaerts (2004), p. 319. See Tomkins (1999), p. 222: for applying there were no "standing restrictions" or requirements to show "just cause"; application must be made "in writing and in sufficiently precise manner", and the institution has one month in which inform the applicant that it intends to refuse the application. Then, the applicant has one month in which to make what the code refers to as "confirmatory application". The code also provides that "if a confirmatory application is submitted, and if the institution concerned decides to refuse to release the document, that decision, which must be made within one month of submission of the confirmatory application, will be notified in writing to the applicant as soon as possible. The grounds for the decision must be given, and the decision must indicate the means of redress that are available, i.e. judicial proceedings and complaints to the ombudsman". As to the series of circumstances that justify the institutions in providing access, the five mandatory reasons regard: the "protection of the public interests" (such as public security, international relations, monetary stability, court proceedings, inspections and investigations); the protection "of the individual and of privacy"; the protection of the commercial and industrial secrecy; the protection of the "Community's financial interests" and "the protection of the confidentiality as requested by the natural or the legal person that supplied the information or as required by the legislation of the Member State that supplied the information". The discretionary reason to refuse access regards the institution's interest in the confidentiality of its proceedings. See Tomkins (1999), p. 223; Harlow (1999), p. 289.

[12] Lenaerts (2004), p. 319.

[13] Lenaerts (2004), p. 319.

[14] Lenaerts (2004), p. 319.

[15] Lenaerts (2004), p. 319.

[16] For this distinction, see more in detail Santini (2004), pp. 131 ff., Alemanno (2014), p. 78 ff.; Sabbatini (2003), p. 391, Vos et al. (2023), p. 49.

Specifically, the doctrine has distinguished between active and passive models of transparency. In *active models* institutions take initiative in providing information to the public without request, for instance, allowing the public to attend the sessions of those institutions.[17] In *passive models* the information is directly requested by the public such in the case of access to documents.[18] Passive transparency entails more in general access to information.[19] As it can be seen, these two models of transparency refer to different institutions.

Active transparency requires the EU institutions to "proactively safeguard the transparency of their proceedings (before, during and after) and outputs".[20] Essentially, the active models of transparency regard the possibility to attend the meetings of the governing bodies. In these models are included occurrences that have received particular attention since the nineties. The publicity of the institution's meetings not only concerns the European Parliament. There has also been a gradual introduction of different forms of publicity for the Council's activities.[21]

Moreover, the interest groups' participation in the decision-making process relates to a different aspect of transparency's active models.[22]

We must recall that the doctrine does not conceive "active" transparency as a fundamental principle of European Union law. Consequentially, scholars put "active transparency" on a "meta-legal" level, as it is expressed in mere practices or in sort

---

[17] Blanke (2013), p. 73.

[18] Blanke (2013), p. 73.

[19] Alemanno (2014), p. 78.

[20] Alemanno (2014), p. 78.

[21] As regard to the Council, the process regarded mainly the exceptions to the principle of the secrecy of deliberations and the opening to the public of Council's meetings. See Santini (2004), pp. 147 ff. As observed by Craig (2018), p. 390, the European Council agreed on an "overall policy on transparency" in June 2006 (see Presidency Conclusion of the Brussels European Council of 15/16 June 2006, available at https://www.consilium.europa.eu/ueDocs/cms_Data/docs/pressData/en/ec/90111.pdf). All Council deliberations on legislative acts to be adopted with the co-decision procedure were to be open to the public, as well as the votes and explanations of votes by Council members. The initial deliberation on legislative acts other than those adopted by co-decision and the debate on Commission's annual work programme were also public. After the Lisbon Treaty, transparency features in the legislative process in Article 16(8) TEU, which imposes on the Council an obligation to meet in public when it deliberates and votes on a draft legislative act (see Craig (2018), p. 391). This is also provided in Article 15(2) TFEU concerning the so-called institutional transparency. Accordingly, the European Parliament and the Council must meet in public when considering and voting on a draft legislative act. Moreover, since documents submitted to the Council open sessions and the relevant minutes are public, open meetings are connected with access to documents (Council Decision No 2009/937/EU adopted the Council's Rules of Procedure). See, Klamert (2019b), pp. 397–404; Goudappel and Van den Brink (2011), pp. 389 ff. See also Hillebrandt et al. (2014), pp. 1–20.

[22] For the topic of the interest groups' participation in the decision-making process, both from the point of view of the regulation of the interest groups' activities ("external") and the regulation aimed at providing guarantees on the members of the institutions subject to the pressure from interest groups ("internal"), see more in detail Santini (2004), pp. 162 ff.

of political commitments.[23] Given the purposes of this book, we will not cover those aspects related to active transparency.

Debates around *recognizing a general principle of transparency* have largely focused on the passive models of transparency and specifically the right of *access*,[24] particularly after the changes introduced by the Treaty of Amsterdam[25] in Article 255 TEC and by the Charter of Fundamental Rights in Article 42 which enshrines the right of access to documents.[26]

Nevertheless, it has been argued that the "uncertain content" of transparency may explain why transparency is frequently connected with other well-established principles of law.[27] This aspect may also influence the perception of transparency as a general principle of European Union law.

The principle of transparency has developed across multiple areas of EU law, from the broadest possible access to documents to the prohibition of altering the tender after the conclusion of a contract.[28] The principle of transparency first appeared in the case law of the Court of Justice in 1995 in *Carvel v. Council*[29] on public access to information. It was then recognised in *Commission v. Belgium*[30] of 1996 in the context of public procurement and further applied in legal cases across different fields.[31] European Courts have played a role in the principle of transparency development because it occurred regularly in their case law and the subject matter varied. For instance, about one third of all the cases decided up until 2008 concerned public access to documents, almost one third was linked to public procurements, and the rest was constituted of cases about the Union's institutional structure, competition and state aid law and other occurrences.[32]

It has also been noted that transparency served as a basis for secondary legislation, such as for the public procurement directives or the directive on the

[23] Santini (2004), p. 274.

[24] Santini (2004), p. 272; Craig (2018), p. 392 ff.

[25] Treaty of Amsterdam amending the Treaty on European Union, the Treaties establishing the European Communities and certain related acts, OJ C 340, 10.11.1997.

[26] Charter of Fundamental Rights of the European Union proclaimed on 7 December 2000, OJ C 202, 7.6.2016.

[27] In this sense, Prechal and de Leeuw (2007), p. 61. See also Addink (2009), p. 239.

[28] Buijze (2013), p. 5.

[29] Judgment of the Court of First Instance (Second Chamber, extended composition) of 19 October 1995, *John Carvel and Guardian Newspapers Ltd v Council of the European Union*, Case T-194/94, ECLI:EU:T:1995:183.

[30] Judgment of the Court (Fifth Chamber) of 25 April 1996, *Commission of the European Communities v Kingdom of Belgium*, Case C-87/94, ECLI:EU:C:1996:161.

[31] See Santini (2004), p. 70.

[32] See Buijze (2013), p. 14: the other occurrences concern telecommunications, VAT, regulation of the energy market, equal pay, consumer law, data protection, Treaty freedoms, language rules, milk quota, notification of Commission decisions and research grants.

establishment of the internal market for telecommunication services through the implementation of open network provisions.[33]

After the Lisbon Treaty, transparency features as a *general objective* of the European Union.[34] Article 1 TEU states that the Lisbon Treaty "marks a new stage in the process of creating an even closer union among the people of Europe, in which decisions are taken as openly as possible and as closely as possible to the citizens".[35] This statement is reiterated in Article 10 (3) TEU, which provides that "every citizen shall have the rights to participate in the democratic life of the Union" and "decisions shall be taken as openly and as closely as possible to citizens", showing the close link to democracy.[36]

The Lisbon Treaty in Article 11 (2) TEU links transparency with dialogue and consultation, mandating the EU institutions to maintain an "open, transparent and regular dialogue" with civil society. Article 11 (3) TEU further requires the Commission to carry out broad consultations with parties in order to ensure that the Union's actions are coherent and transparent. Another fundamental provision is Article 15 (1) TFEU which mandates the institutions to conduct their work "as openly as possible" in order to promote good governance and participation.[37] Finally, after the Lisbon Treaty the European Union made a number of initiatives to enhance transparency, such as the Commission Decisions of 2014 on the publication of information on meetings and Commission's new rules on expert groups in 2016.[38] Among the key focus of the European Council's strategic agenda for 2019–2024, for instance, appeared the principle of democracy, the rule of law and transparency, while each institution should "revisit its working methods and reflect on the best way to fulfil its role under the Treaties".[39]

Initially, legal doctrine was reluctant to recognise *transparency as a principle of law*. Some scholars fail to mention the principle of transparency,[40] despite the widening of the scope of the obligations connected to transparency and the establish-

[33] See Council Directive 90/387/EEC of 28 June 1990. See Buijze (2013), pp. 6–7. The Author also recalls other examples, such as Directive 89/105/EEC of 21 December 1989 relating to the transparency of measures regulating the pricing of medicinal products for human use and their inclusion within the scope on national health insurance system and the Council Directive 90/684/EEC of 21 December 1990 on aid to shipbuilding which included "a principle of transparency of aid for shipbuilding and ship conversion".

[34] Craig (2018), p. 391.

[35] See also Blanke (2013), pp. 71–74.

[36] Klamert (2019a).

[37] See also Craig (2018), p. 391, Volpato et al. (2023), pp. 371–373, and Vanuzzo (2024), p. 179.

[38] Commission Decision of 25 November 2014 on the publication of information on meetings held between Directors-General of the Commission and organisation of self-employed individuals (C(2014) 9048) and Commission Decision of 25 November 2014 on the publication of information on meetings held between Members of the Commission and organisation or self-employed individuals (C(2014) 9051); Commission Decision C(2016) 3301 of 30 May 2016 establishing horizontal rules on the creation and operation of Commission expert groups. See Craig (2018), p. 392.

[39] European Council (2019). See Hoffman and Leino-Sandberg (2019).

[40] Tridimas (2006), p. 19.

ment of transparency as a principle linked to public procurement.[41] Others, to which the majority legal doctrine adheres, consider access to documents a self-standing principle or even a general principle of EU law.[42]

Hence, we should briefly recall which requirements a principle needs to be identified as a general principle of EU law. General principles are recognised as one of the sources of EU law, "based" on Article 19 (1) TEU, and play an important role in legal reasoning.[43] In the doctrine, a principle is considered general when it incorporates a minimum ascertainable legally binding content and it enjoys a degree of wide acceptance.[44] A general principle also *transcends* specific areas of law and underlies a legal system as a whole.[45] Such a principle should be justifiable at least "as a standard of review of the Union's acts".[46]

A more integrative approach toward the principle of transparency emerged in those who argue that transparency is an important legal concept and it includes access to information, the right of access to file, the right to be heard before a decision is taken, the right to receive a reason for a decision.[47]

Although transparency is deemed to be *linked to other principles*, some assert that there are different "open-ended" elements of transparency that might be viewed as "building blocks" of an "overarching" principle.[48] For instance, it has been noted that it is possible to identify a number of "recurring core elements of transparency",

[41] Bovis (2009), p. 289 ff. affirms that transparency is a principle of public procurement law.

[42] Craig (2018), p. 388 ff.; Craig and De Búrca (2020), p. 602 ff.

[43] Buijze (2013), p. 71. See Tridimas (2006), p. 19 (the author refers to the previous Article 220 EC, now 19 TEU). See Semmelmann (2013), pp. 457–487; Von Bogdandy (2010), pp. 35–56. See also Neuvonen and Ziegler (2022), pp. 7–24.

[44] Tridimas (2006), p. 26. We recall the author's words: "as a general guidance, the principle must incorporate a minimum ascertainable legally binding content. In the absence of guidance by Community written law, it must be widely accepted in one way or another by the Member States. […] To be elevated to the status of a general principle, a preposition must enjoy a degree of wide acceptance, i.e. represent 'conventional morality'".

[45] Tridimas (2006), pp. 1 ff.: "As a starting point, it may be said that a principle is a general preposition of law of some importance from which concrete rules derive. According to this definition, the constituent elements of a principle are two: it must be general and it must carry added weight. 'General' means that it operates at a level of abstraction that distinguishes it from a specific rule. The term 'general' may have other meanings. It may refer to principles which transcend specific areas of law and underlie the legal system as a whole. It may also refer to the degree of recognition or acceptance. In that sense, an unwritten legal postulate cannot enjoy the status of a general principle unless it can somehow objectively be verified that it enjoys a minimum degree of recognition by a relevant constituency, e.g. the courts, or political actors, or the citizenry, or the constituent members of a supra-national legal order. The second element of the definition is that the principle must be of some importance. It must express a core value of an area of law or the legal system as a whole. There are clearly degrees of importance. Thus, in a narrower sense, the term general principle may be reserved for fundamental propositions of law which underlie a legal system and from which concrete rules or outcomes may be derived".

[46] Karageorgou (2012), p. 15; De Witte (2000), pp. 143–162.

[47] Vesterdorf (1998), pp. 902 ff.; Santini (2004), Buijze (2013), Harlow (1999), Tomkins (1999), p. 255, Goudappel and Van den Brink (2011).

[48] Prechal and de Leeuw (2007), p. 52.

such as clear language, physical access to information, publication or notification, the predictability of public authorities' actions, and "consistency in the interpretation and application of the law".[49] However, these aspects are not analysed in-depth by scholars, who generally confirm that only access to documents may be viewed as a general principle.

Given this context, two main aspects are particularly useful for a better understanding of the principle of transparency in EU law.

The first relates to the idea that transparency may be implicit and inherent in other legal principles. The overlaps of transparency with other legal principles concern mainly some other well-established EU law principles.

The second focuses on the pivotal topic of access to documents. We will briefly describe the context in which access to documents arises and its legislative and judicial developments. It seems particularly important to approach this issue, as access to documents is the only "serious nominee"[50] to be accepted as a general principle of EU law according to the prevailing doctrine.

3.2 Relationships and Overlaps with Other Legal Principles

Transparency emerged on so many levels next to other legal principles or—occasionally—*rather than an established principle.*[51] The doctrine described transparency as a concept linked to the principle of sound administration, the principle of legal certainty, the principle of equal treatment and the principle of sincere cooperation. It is useful to recall these connections in order to understand how the CJEU and the doctrine qualify transparency in this area.

3.2.1 Sound Administration

It has been argued that transparency is linked to the principle of sound administration. The right of sound administration is one of the common general principles of the Member States' constitutional tradition and it is observed in a State governed by the rule of law.[52] It can be viewed a sort of "closing principle" on relations between European institutions and citizens.[53] In this regard, it is worth mentioning that Article 41 of the Charter of Fundamental Rights enshrines the principle of sound

[49] Prechal and de Leeuw (2007), p. 61.

[50] Prechal and de Leeuw (2007), p. 61.

[51] Prechal and de Leeuw (2007), p. 61.

[52] Lenaerts (2004), p. 336.

[53] Santini (2004), p. 119.

administration.[54] Particularly, Article 41 (1) states that "every person has the right to have his or her affairs handled impartially, fairly and within a reasonable time by the institutions, bodies, offices and agencies of the Union". In addition, Article 41 provides in paragraph 2 that the right includes "the right of every person to be heard, before any individual measure which would affect him or her adversely is taken; the right of every person to have access to his or her file, while respecting the legitimate interests of confidentiality and of professional and business secrecy; the obligation of the administration to give reasons for its decisions".[55]

The principle of sound administration "represents in fact an overarching notion encompassing various legal concepts, overlapping with other procedural fundamental rights".[56]

In the Court of Justice case-law this aspect arose, for instance, with regard to the *Natural Health* case.[57] Interestingly, this case refers to the legitimacy of the procedure under the Directive 2002/46 concerning food supplements and in particular the authorisation procedure of certain supplements, such as vitamins or minerals, to be placed on the market. After the authorisation, the supplements are included in the so-called positive list.

According to the Advocate General in *Natural Health*, this authorisation procedure "has the *transparency of a black box*: no provision is made for parties to be heard, no time-limits apply in respect of decision-making; nor, indeed, is there any certainty that a final decision will be taken".[58] In the Advocate general's opinion, the Directive lacked appropriate "transparent" procedures for its applications.[59] Therefore, it deemed the authorisation procedure invalid. Although the Court of Justice found that the procedure was legal, it pointed out that "it would, no doubt, have been desirable, as regards the stage between the filing of a dossier seeking modification of the positive lists and the time when the matter is brought before the committee (a stage which includes, inter alia, consultation of the European Food Safety Authority as envisaged in both Article 14 of, and the 10th recital to, Directive 2002/46/EC), for the directive to have included provisions which in themselves ensured that that stage be *completed transparently* and within a reasonable time".[60]

---

[54] Lock (2019a), p. 2204 ff.

[55] Lock (2019a), p. 2204 ff. See also Weiss and Steiner (2007), pp. 1551–1554.

[56] Lenaerts (2004), p. 336.

[57] Judgment of the Court (Grand Chamber) of 12 July 2005, *The Queen, on the application of Alliance for Natural Health and Nutri-Link Ltd v Secretary of State for Health* (C-154/04) and *The Queen, on the application of National Association of Health Stores and Health Food Manufacturers Ltd v Secretary of State for Health and National Assembly for Wales* (C-155/04), Joined cases C-154/04 and C-155/04, ECLI:EU:C:2005:449.

[58] Opinion of Advocate General Geelhoed delivered on 5 April 2005, para 85. Emphasis added. On the authorisation procedures of the food supplements, Van der Meulen (2020), p. 214.

[59] In paragraph 85 the AG continues: "the procedure therefore lacks essential guarantees for the protection of the interests of private applicant". See Prechal and de Leeuw (2007), p. 53.

[60] Judgment of the Court (Grand Chamber) of 12 July 2005, Joined cases C-154/04 and C-155/04, para 81. Emphasis added.

The Court added that the lack of those provisions should be compensated by the Commission that had to adopt "[...] in accordance with the *principle of sound administration*, the measures necessary to ensure generally that the consultation stage with the European Food Safety Authority is carried out transparently and within a reasonable time".[61] Therefore, this case law demonstrates that transparency has no independent value, but it is linked to the principle of sound administration which directs the institutions' actions. Nevertheless, in this principles-oriented interpretation the Court's argument also draws upon the *Meroni* principle as part of the broader principle of institutional balance in EU law, which is still leading when it comes to "restrictions that must be placed on 'outsourced' executive competences".[62]

In a similar way, the principle of sound administration can be the basis of institutions' rules of conduct that cannot be derived in any other way.

The Court of Justice ascribed to this principle the requirement for the institutions to disclose information on the criteria that had determined the content of certain acts. For instance, the CJEU has indicated a possible use of the principle of sound administration in order to overcome the informational limits inherent in the duty to give reasons.[63] The duty to give reasons constitutes a specific transparency obligation because it requires public authorities to communicate the reasons of their decisions.[64] In the *Tradax* case the Court affirmed that "in the statement of the reasons on which a measure fixing free-at-frontier prices is based, the administration is entitled to confine itself to setting out in a general form the essential factors and the procedure which formed the background to its evaluation of the facts, without its

---

[61] Judgment of the Court (Grand Chamber) of 12 July 2005, Joined cases C-154/04 and C-155/04, para 82. Emphasis added. See also Vandamme (2006), p. 313 ff. According to the author (at p. 314), "the shortcomings of the Directive in this regard are being smoothened out by a Treaty-consistent or rather a principle-consistent interpretation of the Directive. The executive powers attributed to the Commission must be seen in the light of the general principles of Community law, in particular the principle of sound administration and the principle of legal certainty. It is therefore up to the Commission to ensure a transparent, orderly and swift procedure before the EFSA. Under this line of argumentation, the CJEU 'saves' the Directive's validity that was in the eyes of the Advocate-General so clearly invalid. Yet, the Court's reasoning can be said to be conflicting with the principle that the most important structural elements of Community legislation must be laid down in secondary Community law itself. It would seem to me that the basic elements of the procedure before the EFSA, possibly leading up to a definitive amendment of the Positive Lists, are part of such structural elements".

[62] Vandamme (2006), p. 314. The Court in Judgment of the Court (Grand Chamber) of 12 July 2005, Joined cases C-154/04 and C-155/04, para 90 states that "[...] when the Community legislature wishes to delegate its power to amend aspects of the legislative act at issue, it must ensure that that power is clearly defined and that the exercise of the power is subject to strict review in the light of objective criteria (see, to that effect, Case 9/56 *Meroni v. High Authority* [1958] ECR 133, at p. 152) because otherwise it may confer on the delegate a discretion which, in the case of legislation concerning the functioning of the internal market in goods, would be capable of impeding, excessively and without transparency, the free movement of the goods in questions". On the *Meroni* judgment, see Chap. 5, para. 5.2.

[63] Santini (2004), p. 120.

[64] Buijze (2013), p. 273.

being necessary to specify the facts themselves".[65] Then it added: "it should however be pointed out that it would be *consistent with good administration* for the Commission periodically to publish for the information of the traders concerned the main data taken into account in fixing cif prices".[66]

Some codes of good administrative behaviour also include the concept of good administration as outlined.[67] As seen in the previous chapter, the Code of Good Administrative Behaviour drafted by the European Ombudsman identifies transparency as a good administration requirement.

Some case laws about transparency and sound administration deal with the state aid domain. According to the rule that decisions must be taken without delay and addressed to the Member State concerned in the interest of transparency, a failure to notify the Commission's decision can justify the annulment of an Institution's act.[68] Despite that, the Commission may also be bound to inform the complainant about its decisions and its consequences in line with "the principle of sound administration and transparency".[69]

3.2.2 Legal Certainty

The connection between transparency and legal certainty encompasses the requirements of clarity and the "unambiguous nature" of legal texts.[70] Clarity and publication of legal texts assure the law's recognisability in such a way that the European legislation is foreseeable and certain for citizens.[71] Legal certainty and transparency share the same purpose of facilitating decision-making by allowing individuals to be aware of the content of law and the consequences of their actions. Even though

---

[65] Judgment of the Court (First Chamber) of 15 March 1984, *Tradax Graanhandel BV v Commission of the European Communities*, C-64/82, ECLI:EU:C:1984:106, para. 21.

[66] Judgment of the Court (First Chamber) of 15 March 1984, *Tradax Graanhandel BV v Commission of the European Communities*, C-64/82, ECLI:EU:C:1984:106, para. 22. Emphasis added, See Santini (2004), p. 120.

[67] As specified by Santini (2004), p. 122, these codes include numerous rules of conduct, some of which merely set out general principles that have long been widely recognised in EU Law, while others appear to be directly based on the principle of sound administration.

[68] Prechal and de Leeuw (2007), p. 53.

[69] Judgment of the Court (Sixth Chamber) of 18 June 2002, *Kingdom of Spain v Commission of the European Communities*, C-398/00, ECLI:EU:C:2002:382; Judgment of the Court of First Instance (Fourth Chamber, extended composition) of 17 June 1999, *Associação dos Refinadores de Açúcar Portugueses (ARAP), Alcântara Refinarias - Açúcares SA and RAR Refinarias de Açúcar Reunidas SA v Commission of the European Communities*. Case T-82/96, ECLI:EU:T:1999:127, Judgment of the Court of First Instance (Second Chamber, extended composition) of 22 May 1996, *Associazione Italiana Tecnico Economica del Cemento (AITEC) v Commission of the European Communities*, Case T-277/94, ECLI:EU:T:1996:66. See Prechal and de Leeuw (2007), p. 53.

[70] See Prechal and de Leeuw (2007), p. 53. See also Groussot et al. (2017), pp. 77–104.

[71] See Buijze (2013), p. 271; Alemanno (2014), p. 82; Prechal and de Leeuw (2007), p. 54.

clarity, precision and publication relate to transparency, in case law they are deemed to pertain to legal certainty and transparency only indirectly.[72]

For instance, in the *Belgium v. Commission case*, the Belgian government sought the annulment of Regulation (EC) No 2204/2002 on aid to employment.[73] It argued that it lacked clarity in terms of content and of context even if the preamble of the enabling regulation (Regulation (EC) No 994/98) requires exemption regulation to increase transparency and legal certainty.[74] According to the Advocate General's opinion, "both the principle of transparency and that of legal certainty must be respected by the legislature as sources of Community law, and a failure to do so would, under Article 230 EC, constitute an infringement, irrespective of whether they are referred to in the preamble to Regulation No 994/98".[75] He concluded affirming that there was no breach of these principles. The AG also specified that "transparency is concerned with the *quality of being clear, obvious and understandable without doubt or ambiguity*. The application of this principle in the field of law is something of an aspiration, as the translation of the law into everyday life is not straightforward and does not always offer clear answers".[76] The Court of Justice, however, refers only to the breach of the general principle of legal certainty. Therefore, it has been argued that the CJEU's approach was to consider transparency only as an element of the principle of legal certainty.[77]

Interestingly, certain rulings of the Court of Justice confirm that the publication of documents exercises a general function of legal certainty. For instance, the Court held that certain conditions to which the use of Parma ham was subject could not be relied on against business operators, since they had not been adequately publicised in the relevant Regulation, but they were laid down in unpublished acts.[78] In the case *Consorzio del Prosciutto di Parma,* the Court stated that "the scope and effect of a Community provision must be clear and foreseeable to individuals, otherwise the principle of legal certainty and the principle of transparency are breached".[79] Therefore, the Court specified that "if the Consorzio were authorised to obtain, before a national court, compliance with an unpublished specification, the

[72] Prechal and de Leeuw (2007), p. 54; Klamert (2019a), p. 21; Karageorgou (2012), pp. 10 ff.

[73] Judgment of the Court (Third Chamber) of 14 April 2005, *Kingdom of Belgium v Commission of the European Communities*, C-110/03, ECLI:EU:C:2005:223.

[74] Prechal and de Leeuw (2007), p. 54.

[75] Opinion of Mr. Advocate General Ruiz-Jarabo Colomer delivered on 16 December 2004, *Kingdom of Belgium v Commission of the European Communities*, ECLI:EU:C:2004:815, para. 36.

[76] Opinion of Mr. Advocate General Ruiz-Jarabo Colomer, para. 44. Emphasis added.

[77] Prechal and de Leeuw (2007), p. 54.

[78] See Judgment of the Court of 20 May 2003, *Ravil SARL v Bellon import SARL and Biraghi SpA*, C-469/00, ECLI:EU:C:2003:295; Santini (2004), p. 114.

[79] Judgment of the Court of 20 May 2003, *Consorzio del Prosciutto di Parma and Salumificio S. Rita SpA v Asda Stores Ltd and Hygrade Foods Ltd.*, Case C-108/01, ECLI:EU:C:2003:296, para 85.

principles of legal certainty and transparency would be breached" and consequently, "the provisions relating to that specification cannot have direct effect".[80]

In another case, the Court did not argue about the possibility of the applicant to rely on the principle of transparency, but it found the decision under scrutiny "clear in every aspect".[81] Similarly, transparency sometimes appears as a requirement separated from the principle of legal certainty in case law on the proper implementation of directives.

For instance, there are cases in which the Court affirmed that provisions of directives must be implemented in national law "with precision, clarity and transparency required in order to comply fully with the requirement of legal certainty", or the provisions "must be capable of creating a situation which is sufficiently precise, clear and transparent to enable individuals to ascertain their rights and obligations".[82]

Besides clarity of legal texts, a second aspect of legal certainty stems from its link with transparency. Precisely, transparency serves as a *safeguard* against the arbitrary behaviour of the institutions by improving the policy actions' predictability.[83] In some case law, the Court has ruled that whenever the Commission has posed a limit on the exercise of its discretionary powers by adopting soft law instruments such as guidelines and rules of conduct producing external effects, it cannot contravene them at will.[84] This aspect can be found in case law on state aid and in competition areas.

In the former, for instance, the Court in case C-310/99 has argued that the guidelines concerning aid to employment helped ensuring that the Commission acted in a foreseeable and transparent manner according to the principle of legal certainty.[85]

In competition cases, after having been criticised for its methods for calculating fines, the Commission adopted guidelines to ensure transparency and impartiality in that area.[86] In the related case law, the CJEU argued that transparency serves to

[80] Judgment of the Court of 20 May 2003, *Consorzio del Prosciutto di Parma and Salumificio S. Rita SpA v Asda Stores Ltd and Hygrade Foods Ltd.*, para 85.

[81] Prechal and de Leeuw (2007), p. 55. The authors refer to Judgment of the Court of 23 November 1999, *Portuguese Republic v Council of the European Union*, C-149/96, ECLI:EU:C:1999:574. See paragraph 55 "the Portuguese Government contends that the *principle of transparency* has been breached because the contested decision approves Memoranda of Understanding which are not adequately structured and are drafted in obscure terms which prevent a normal reader from immediately grasping all their implications, in particular as regards their retroactive application. In support of this plea, it relies on the Council Resolution of 8 June 1993 on the quality of drafting of Community legislation".

[82] Prechal and de Leeuw (2007), p. 55. The authors refer to Judgment of the Court (Fifth Chamber) of 13 September 2001, *Commission of the European Communities v Kingdom of Spain*, C-417/99, ECLI:EU:C:2001:445, in particular at para. 40.

[83] Prechal and de Leeuw (2007), p. 54.

[84] Prechal and de Leeuw (2007), p. 56.

[85] Judgment of the Court (Sixth Chamber) of 7 March 2002, *Italian Republic v Commission of the European Communities*, C-310/99, ECLI:EU:C:2002:143, recalled by Prechal and de Leeuw (2007), p. 55.

[86] Prechal and de Leeuw (2007), p. 56.

ensure legal certainty by making public the criteria for setting the fines.[87] Therefore, from the point of view of legal certainty, transparency in the conduct of institutions is essential. Namely, they must disclose the legal and factual context in which the adoption of certain measures may take place so that these measures may be foreseeable. This requirement also appears in the *AKZO II case*. With reference to a decision by which the Commission authorised its member responsible for competition matters to adopt certain procedural measures provided for in Regulation 17/62, the Court of Justice observed that "the principle of legal certainty and the need for administrative decisions *to be transparent* require that the Commission should publish decisions granting delegations of authority as if they were rules of procedure (…) which lays down the general framework in which such decisions are adopted".[88] The Court of Justice did not draw any practical consequences from its statements. It dismissed the action brought before it, finding that the non-publication of the authorisation decision did not deprive the applicants of the possibility of challenging its regularity or contesting the decision taken on the basis of the authorisation thus granted.[89]

However, it has been noted that in these cases, transparency functions as a "mechanism to prevent arbitrary behaviour on the part of the institution in question".[90] Therefore, it differs from transparency as a requirement of clarity and publication of legal texts.

3.2.3 *Equal Treatment*

The relationship between equal treatment and transparency has been mostly studied in the context of non-discrimination in public procurement. It has been reported that one of the first judgments about transparency occurred in the case C-87/94 *Commission v. Belgium*, in which the Court held on the basis of the text of the

[87] Prechal and de Leeuw (2007), p. 56.

[88] Judgment of the Court (Fifth Chamber) of 23 September 1986, *AKZO Chemie BV and AKZO Chemie UK Ltd v Commission of the European Communities*, C-5/85, ECLI:EU:C:1986:328, in particular para. 39. Emphasis added.

[89] Santini (2004), p. 29, pp. 115 ff. The author mentions also Judgment of the Court (First Chamber) of 22 April 1999, *Kernkraftwerke Lippe-Ems GmbH v Commission of the European Communities*, C-161/97, ECLI:EU:C:1999:193, in the EAEC context in which a German company owning a nuclear power plant challenged the Commission's decision confirming a decision of the Supply Agency established by the EAEC Treaty prohibiting that company from importing uranium from the CIS. In its appeal, the company claimed, inter alia, that the principle of legal certainty had been infringed because, in its opinion, the Agency had acted in a non-transparent manner. The Court of Justice did not contest the functional link between the transparency of the Agency's conduct and legal certainty. It therefore dismissed the appeal.

[90] Prechal and de Leeuw (2007), p. 57. In the context of the relationship between transparency and legal certainty, Santini (2004), p. 114, pp. 261 ff. recalls the further aspect of the quality of drafting (see, more in detail, pp. 261 ff.).

Directive 90/531, that the procedure for comparing tenderers had to respect the principle of equal treatment of the tenderers and the principle of transparency.[91] Precisely, the Court stated that "the procedure for comparing tenders therefore had to comply at every stage with both the principle of the equal treatment of tenderers and the principle of transparency so as to afford equality of opportunity to all tenderers when formulating their tenders".[92]

It has been noted that the CJEU developed the relationship between transparency and non-discrimination more thoroughly in a number of cases concerning public procurement and concessions. The Court of Justice has indeed pointed out that the principle of equal treatment concerns not only public procurement but also free movement rules of the internal market which govern, among others, the award of concessions.[93] Thus, in the field of public procurement and concessions, the link between transparency and non-discrimination is established and has its most important practical applications, both within the Member States (which in particular are obliged to implement the directives on public procurement of works, services and supplies) and in relation to the invitations to tender issued by the institutions.[94]

It is, therefore, a constant affirmation of the Court of Justice that the principle of non-discrimination on the grounds of nationality implies, in particular, an obligation of transparency in order to ascertain its compliance.[95] It "consists in ensuring, for the benefit of any potential tenderer, a degree of advertising sufficient to enable the services market to be opened up to competition and the impartiality of procurement procedures to be reviewed".[96]

---

[91] Judgment of the Court (Fifth Chamber) of 25 April 1996, *Commission of the European Communities v Kingdom of Belgium*, C-87/94, ECLI:EU:C:1996:161. See Prechal and de Leeuw (2007), p. 57; Buijze (2013), p. 6; Santini (2004), p. 119.

[92] *Commission of the European Communities v Kingdom of Belgium*, C-87/94, para. 54.

[93] Prechal and de Leeuw (2007), p. 57. They refer to Judgment of the Court (Sixth Chamber) of 12 December 2002, *Universale-Bau AG, Bietergemeinschaft: 1) Hinteregger & Söhne Bauges.m.b.H. Salzburg, 2) ÖSTÜ-STETTIN Hoch- und Tiefbau GmbH v Entsorgungsbetriebe Simmering GmbH*, C-470/99, ECLI:EU:C:2002:746, Judgment of the Court (Grand Chamber) of 21 July 2005, *Consorzio Aziende Metano (Coname) v Comune di Cingia de' Botti*, C-231/03, ECLI:EU:C:2005:487, Judgment of the Court (First Chamber) of 13 October 2005, *Parking Brixen GmbH v Gemeinde Brixen and Stadtwerke Brixen* AG, C-458/03, ECLI:EU:C:2005:605.

[94] See Santini (2004), p. 118. The author recalls that in the context of the ECSC, case law had considered the "instrumental link between publicity and non-discrimination" rather than transparency (see, on this point, p. 117 and the reference to Judgment of the Court (Fourth Chamber) of 10 July 1985, *Wirtschaftsvereinigung Eisen-und Stahlindustrie v Commission of the European Communities*, Case 27/84, ECLI:EU:C:1985:311).

[95] Santini (2004), p. 118, Judgment of the Court (First Chamber) of 18 November 1999, *Unitron Scandinavia A/S and 3-S A/S, Danske Svineproducenters Serviceselskab v Ministeriet for Fødevarer, Landbrug og Fiskeri*, C-275/98, ECLI:EU:C:1999:567, particularly at para. 31: "It should be noted, however, that the principle of non-discrimination on grounds of nationality cannot be interpreted restrictively. It implies, in particular, an obligation of transparency in order to enable the contracting authority to satisfy itself that it has been complied with".

[96] Judgment of the Court (Sixth Chamber) of 7 December 2000, *Telaustria Verlags GmbH and Telefonadress GmbH v Telekom Austria AG*, Case C-324/98, ECLI:EU:C:2000:669, para. 62.

Furthermore, the principle of transparency in the case *Succhi di Frutta* is considered as the "corollary" of the principle of equal treatment between tenderers.[97]

The principle of equal treatment is deemed to contain at least two obligations of transparency.

The first regards the "creation of equal opportunities" placing all potential bidders on a level playing field.[98] Accordingly, transparency gives to all interested parties equal opportunities "in formulating the terms of the applications for and participation in the tenders".[99]

The second reason why the principle of equal treatment implies an obligation of transparency is to "facilitate the compliance's control".[100] According to the CJEU, transparency enables contracting or awarding authorities to ensure compliance with the principles of equal treatment and non-discrimination and the impartiality of procurement procedures.[101] Such control can be possible either during the award procedure or *ex post*, so that tenderers and authorities can verify whether the principle of equal treatment has been observed.[102]

In this context, the fact that transparency is a way to assess whether non-discrimination has been observed seems to extend in some respect transparency *beyond* the principle of equal treatment.[103] Transparency can therefore be separated from equal treatment. Furthermore, since equality of treatment underpins the Treaty

---

[97] Judgment of the Court (Sixth Chamber) of 29 April 2004, *Commission of the European Communities v CAS Succhi di Frutta SpA*, Case C-496/99, ECLI:EU:C:2004:236, see Prechal and de Leeuw (2007), p. 58; Santini (2004), p. 119. The author recalls that the Court of First instance in Judgment of 14 October 1999 *CAS Succhi di Frutta SpA v Commission of the European Communities* Joined cases T-191/96 and T-106/97, ECLI:EU:T:1999:256 made clear in para. 73 that "the Commission was obliged to specify clearly in the notice of invitation to tender the subject-matter and the conditions of the tendering procedure, and to comply strictly with the conditions laid down, so as to afford equality of opportunity to all tenderers when formulating their tenders. In particular, the Commission could not subsequently amend the conditions of the tendering procedure, and in particular those relating to the tender to be submitted, in a manner not laid down by the notice of invitation to tender itself, without offending against the principle of transparency".

[98] Prechal and de Leeuw (2007), p. 59.

[99] Prechal and de Leeuw (2007), p. 59. See also Judgment of the Court (Grand Chamber) of 21 July 2005, *Consorzio Aziende Metano (Coname) v Comune di Cingia de' Botti*, C-231/03, ECLI:EU:C:2005:487, para. 17 and 18.

[100] Prechal and de Leeuw (2007), p. 59.

[101] Judgment of the Court (First Chamber) of 13 October 2005, *Parking Brixen GmbH v Gemeinde Brixen and Stadtwerke Brixen* AG, C-458/03, ECLI:EU:C:2005:605, para 49. See Prechal and de Leeuw (2007), p. 59.

[102] Prechal and de Leeuw (2007), p. 59. See also Judgment of the Court (First Chamber) of 11 January 2005, *Stadt Halle and RPL Recyclingpark Lochau GmbH v Arbeitsgemeinschaft Thermische Restabfall- und Energieverwertungsanlage TREA Leuna,* C-26/03, ECLI:EU:C:2005:5, at para 39: "The obligation of transparency, to which the contracting authority is subject in order to make it possible to verify that the Community rules have been complied with [...] should be noted in this respect".

[103] Prechal and de Leeuw (2007), p. 60, Addink (2009), p. 239.

freedoms, scholars argue that it could be a tool to extend transparency requirements, as it occurs in the concession area or in the Services Directive.[104]

3.2.4 *Sincere Cooperation*

The duty of sincere cooperation operates as a fundamental regulating principle in the relations between Community institutions and Member States, but it also imposes mutual obligations of loyal cooperation between the different European institutions.[105]

This principle implies, *inter alia*, duties of information. The Court of Justice has made it clear that this principle not only applies to the transmission of information from the authorities of the Member States to the European institutions, but also in the opposite direction.[106] The "sincere cooperation doctrine" imposes on Member States' administrations and Courts "a duty to act in good faith and to cooperate with the EU institutions" in order to avoid "jeopardising the application of EU law".[107]

---

[104] Prechal and de Leeuw (2007), p. 60. The authors refer to two Directives in which transparency has been codified alongside the requirements of equal treatment and non-discrimination: Directive 2004/18 (procedures for the award of public works contracts, public supply contracts and public service contracts), OJ 2004 L 134/114, in part. Article 2 and Directive 2004/17 (procurement procedures of entities operating in the water, energy, transport and postal services sectors), OJ 2004 L 134/1, in part. Article 10. Moreover, they recall directives concerning liberalization of network sectors, such as Article 7(3) of Directive 2002/20 (authorisation of electronic communications networks and services) (OJ 2002 L 108/21); Article 6 of Directive 2003/54 (common rules for the internal market in electricity) (OJ 2003 L 176/37); Articles 9 and 11 of Directive 2002/21 (common regulatory framework for electronic communications networks and services—the Framework Directive) (OJ 2002 L 108/33) or Article 9 Directive 97/67 (internal market for Community postal services) (OJ 1998 L 15/14). As regards the Services Directive (Directive 2006/123, services in the internal market, OJ 2006 L 376/36), they point out that "In so far as this Directive allows for authorisation schemes, justified by overriding reasons relating to public interest (Article 9), the latter must 'be based on criteria which preclude the competent authorities from exercising their power of assessment in an arbitrary manner.' (Article 10). According to Section 2 of Article 10, these criteria shall, *inter alia*, be '(a) non-discriminatory; ... (d) clear and unambiguous; ... (f) made public in advance; (g) transparent and accessible.' Interesting in the Service Directive is also Article 12, which deals with a limited number of authorisations being available due to the scarcity of natural resources or technical capacity. The Member States are bound to apply a selection procedure to potential candidates which 'provides full guarantees of impartiality and transparency, including, in particular, adequate publicity about the launch, conduct and completion of the procedure.' Article 13(1) provides that authorisation procedures and formalities shall be 'clear, made public in advance and be such as to provide the applicants with a guarantee that their application will be dealt with objectively and impartially" (p. 60).

[105] The current legal basis for the principle of sincere cooperation is Article 4(3) of the Treaty on the European Union (TEU), which provides that "pursuant to the principle of sincere cooperation, the Union and the Member States shall, in full mutual respect, assist each other in carrying out tasks which flow from the Treaties". See Klamert (2014).

[106] Santini (2004), p. 123.

[107] Rovetta and Gambardella (2014), p. 339.

Nevertheless, such doctrine also works the other way around. Thus, EU institutions must cooperate with Member States authorities and national Courts when they seek assistance with regard to issues covered by EU law. For instance, in case *Zwartveld* the CJEU ruled that the Commission must disclose and produce documents requested by national Courts dealing with the application of EU law.[108] In the aforementioned case, the Court established that the Commission should transmit the required documents to the national judicial authority, unless "it presents to the Court imperative reasons relating to the need to avoid any interference with the functioning and independence of the Communities justifying its refusal to do so"[109] or in case of "legitimate grounds connected with the protection of the rights of third parties".[110]

The principle of loyal cooperation has therefore significant consequences for the mutual transparency of European institutions and Member State authorities. This may result in greater transparency of their activities and decision-making processes also by private parties, if the latter have access to the information exchanged between the European and national authorities.[111]

---

[108] Rovetta and Gambardella (2014) p. 339. See Order of the Court of 13 July 1990, J. J. Zwartveld and others, C-2/88 (so-called Zwartveld I) and Order of the Court of 6 December 1990, J. Zwartveld and Others, C-2/88 (Zwartveld II); Santini (2004) p. 123. The Court in *Zwartveld I* at para. 18 and 19 specifies that "this duty of sincere cooperation imposed on Community institutions is of particular importance vis-à-vis the judicial authorities of the Member States, who are responsible for ensuring that Community law is applied and respected in the national legal system. When analysed in the light of those principles, the privileges and immunities which the Protocol grants to the European Communities have a purely functional character, inasmuch as they are intended to avoid any interference with the functioning and independence of the Communities".

[109] Order of the Court of 13 July 1990, *J. J. Zwartveld and others*, C-2/88 (*Zwartveld I*), para. 25.

[110] Order of the Court of 6 December 1990, *J. Zwartveld and Others*, C-2/88 (*Zwartveld II*), para. 11.

[111] Santini (2004), p. 125. The author also recalls the case *Postbank* (Judgment of the Court of First Instance (First Chamber, extended composition) of 18 September 1996, *Postbank NV v Commission of the European Communities*, Case T-353/94, ECLI:EU:T:1996:119) related to cooperation between the Commission and national courts in competition matters in which the Court stated (at para. 69): "that prohibition is not necessary to protect confidential information and business secrets. Once such documents from the administrative procedure are produced in national legal proceedings, there is a presumption that the national courts will guarantee the protection of confidential information, in particular business secrets, since, in order to ensure the full effectiveness of the provisions of Community law in accordance with the principle of cooperation laid down in Article 5 of the Treaty, these authorities are required to uphold the rights which those provisions confer on individuals".

3.3 Transparency and Access to Documents in EU Law

3.3.1 Introduction

The genesis of the principle of transparency in EU law is intrinsically linked to the growing citizens' distrust in governments and to the democratic deficit that occurred in the 1990s, as seen above. After the adoption of the Code of Conduct of 1993 and the approval of the Commission, Council and Parliament's decisions,[112] a further step occurred with the Amsterdam Treaty. In 1997, the Treaty of Amsterdam enshrined the principle of access to documents into primary law, assigning to it the value of a citizens' right in Article 255 TEC. Specifically, Article 255 (1) TEC stated that "any citizen of the Union, and any natural or legal person residing or having its registered office in a Member State, shall have a right of access to European Parliament, Council and documents". Article 255 (2) TEC provided that the general principles concerning such access and its limits should be determined by the Council under the co-decision procedure within two years of the entry into force of the Amsterdam Treaty. Article 255 (3) TEC stipulated that the institutions shall adopt Rules of Procedure concerning access to documents. In addition, Article 42 of the Charter of Fundamental Rights establishes the right of access to documents elevating it as a fundamental right.[113] The Lisbon Treaty in Article 15(3) TFEU (replacing Article 255 TEC) grants a general right of "access to documents of the institutions, bodies, offices and agencies of the EU". The duty to implement the right of access and determine "general principles and limits on grounds of public or private interests" is conferred to the European Parliament and the Council acting pursuing the ordinary legislative procedure.[114] This constitutes the legal basis on which has been adopted Regulation (EC) No 1049/2001 ('Access Regulation') implementing the right of access.[115]

As will be seen further, in the environmental law domain, the Aarhus Regulation (EC) No 1367/2006[116] implementing the Aarhus Convention[117] applies the Access

[112] See Chap. 2, para. 2.2.1.

[113] See Lock (2019b), p 2208: the difference with Article 11 CFR on the right to receive information is that "Article 11 gives a right to access information in the public domain, whereas Article 42 CFR gives access to documents that are often not in the public domain".

[114] See Labayle (2013), p. 11. According to the author, the Treaty enhances the "systematic nature of the triangle of openness/transparency/document access".

[115] Regulation (EC) No 1049/2001 of the European Parliament and of the Council of 30 May 2001 regarding public access to European Parliament, Council and Commission documents, OJ L 145, 31.5.2001.

[116] Regulation (EC) No 1367/2006 of the European Parliament and of the Council of 6 September 2006 on the application of the provisions of the Aarhus Convention on Access to Information, Public Participation in Decision-making and Access to Justice in Environmental Matters to Community institutions and bodies, OJ L 264, 25.9.2006.

[117] Convention on Access to Information, Public Participation in Decision-Making and Access to Justice in Environmental Matters, Aarhus, Denmark, 25 June 1998.

Regulation to requests for *access to environmental information* detained by the institutions.[118] It has been further argued that the European Ombudsman played a key role in the development of transparency as a legal principle and has provided guidance on access to information via the Code of Good Administrative Behaviour.[119]

3.3.2 Access as a Fundamental Right of EU Law

Before the adoption of Regulation (EC) No 1049/2001, the Court of Justice refrained from articulating a general principle of access to documents.[120] Despite that, the Court of Justice played a role in the development of the content of the right of access.

For instance, the Court of Justice annulled a number of decisions of the Commission or the Council refusing access to documents not on the grounds that the institutions "had breached a general principle of transparency", but on different grounds, such as the refusal to consider partial access, the automatic application of the exceptions or the lack or inadequate reasons given.[121]

As previously noted, the first case law on access was *Carvel.*[122] In that case, the CFI stressed that the Council should balance the citizens' interest in obtaining access and the need to maintain confidentiality of its deliberations in deciding whether to release access to documents.[123] Therefore, it shall not automatically refuse entire classes of documents.[124]

In *Netherlands v Council*, the CJEU ruled that there had been a *progressive affirmation* of individuals' right of access to documents held by public authorities "which has been reaffirmed at Community level on various occasions, in particular in Declaration 17 on the right of access to information annexed to the Final Act of the Treaty on European Union ("TEU") which links that right with the democratic nature of the institutions".[125] In this judgment, the Court hasn't pronounced to the extent to identify the right of access as a *general principle* of the legal system, but it left open the possibility of reaching such an outcome in the future. However, the

---

[118] Article 3 Regulation (EC) No 1367/2006.

[119] See Craig (2018), p. 393. See also Tomkins (1999), pp. 235 ff.; Goudappel and Van den Brink (2011), p. 387; Diamandouros (2006).

[120] Craig (2018), p. 393; Weiss and Steiner (2007), p. 1565. See also Harden (2009), pp. 239–256; Heliskoski and Leino (2006), pp. 735–781; Adamski (2012), pp. 521–558.

[121] See Craig (2018), p. 394.

[122] Judgment of the Court of First Instance (Second Chamber, extended composition) of 19 October 1995, *John Carvel and Guardian Newspapers Ltd v Council of the European Union*, Case T-194/94, ECLI:EU:T:1995:183. See para. 3.1.

[123] Craig (2018), p. 393; Santini (2004), p. 226.

[124] On this and other CJEU's rulings on access, see also Lodge (2003), p. 106 ff.

[125] Judgment of the Court of 30 April 1996, *Kingdom of the Netherlands v Council of the European Union*, C-58/94, ECLI:EU:C:1996:171, para 35. Emphasis added. See also Chap. 2, para. 2.2.1. on Advocate General Tesauro's conclusions on the connection between transparency and democracy.

Court hasn't excluded the *recognition* of the right of access also in the regime prior to Article 255 TEC where the right of access was based only on the internal regulations of the institutions.[126]

These two judgments—*Carvel* and *Netherlands v Council*—constitute "the judicial foundation of the emerging law on transparency".[127]

However, in subsequent cases the jurisprudence seems more inclined to elevate the right of access to a general principle of the system.

In *Hautala*[128] the CFI referred to "the principle of the right to information" in a way that seemed to indicate that it considered this principle "in fundamental terms" linked to its democratic origins.[129] In the subsequent appeal, Advocate General Léger emphasized that the right of access "is derived from the most essential political foundations of the Member States of the Community"[130] and proposed to the Court the recognition of the existence of a fundamental right of access to information held by the institutions. Although the CJEU upheld the CFI and endorsed most of the Advocate General's arguments, it avoided to rule on the recognition of a specific principle of EU law.[131]

Even though the Court is "apparently not ready"[132] to acknowledge the existence of a general principle of access to documents, the doctrine points out that *access can be included in the fundamental rights of EU law* pursuant to Article 6 (3) TEU. It states that "fundamental rights, as guaranteed by the European Convention for the Protection of Human Rights and Fundamental Freedoms and as they result from the constitutional traditions common to the Member States, shall constitute general principles of the Union's law".[133] The fact that most Member States' legal systems

---

[126] Santini (2004), p. 196. The author recalls the case WWF UK (Judgment of the Court of First Instance of 5 March 1997, *WWF UK (World Wide Fund for Nature) v Commission of the European Communities*, T-105/95, ECLI:EU:T:1997:26, para 55 in which the CFI has referred that "Although Decision 94/90 is, in effect, a series of obligations which the Commission has voluntarily assumed for itself as a measure of internal organization, it is nevertheless capable of conferring on third parties legal rights which the Commission is obliged to respect". See also Tomkins (1999), pp. 227 ff.; Harlow (1999), pp. 298 ff.

[127] Tomkins (1999), p. 226.

[128] Judgment of the Court of First Instance (First Chamber) of 19 July 1999, *Heidi Hautala v Council of the European Union*, T-14/98, ECLI:EU:T:1999:157.

[129] Curtin (2000), p. 13; Santini (2004), p.198; Weiss and Steiner (2007), p. 1565.

[130] Opinion of Mr. Advocate General Léger delivered on 10 July 2001, *Council of the European Union v Heidi Hautala*, C-353/99, ECLI:EU:C:2001:392, para. 47. According to Tomkins (1999), p. 233, "here we have the first clear judicial statement recognising a legally enforceable principle of a right to information in the Community legal order which goes beyond and is not limited to the provisions of the Commission's and the Council's code of conduct".

[131] Lenaerts (2004), p. 320; Leino (2002), p. 629 ff. In Hautala the CJEU held that the right of access includes access to information contained in the document. The CJEU upheld the CFI's decision to annul the Council's refusal to "consider granting partial access to politically sensitive document" and it held that proportionality required to consider partial access. See Craig (2018), p. 393.

[132] Lenaerts (2004), p. 320.

[133] See Santini (2004), pp. 200 ff; Lenaerts (2004), p. 324.

enshrine the right of access to documents in their constitutions or legislation is considered a sufficient basis to qualify this right as a general EU principle.[134]

The same conclusion can also be reached through the other source referenced in Article 6 (3) TEU. Namely, the fundamental rights enshrined in the European Convention of Human Rights (ECHR).

The right of the public to have access to documents held by authorities would be a necessary corollary of Article 10 ECHR. Article 10 states that "everyone has the right to freedom of expression", which "shall include freedom to hold opinions and to receive and impart information and ideas *without interference* by public authority and regardless of frontiers". Therefore, Article 10 ECHR providing the right to freedom of expression includes also the *freedom to receive information*. As mentioned, the ECHR does not provide a right of access to documents and remains vague on that right.[135]

Nevertheless, through the creative interpretation of the freedom to receive information it is not impossible to construe such a right.[136] Currently, the European Court of Human Rights (ECtHR) seems to override its previous argument made in case *Leander v Sweden* where it denied the existence of an individual right to access to documents at all under the ECHR.[137] Consequently, the ECtHR seems to be in the process of recognising a right of access to official documents in the direction of reading all obligations in Article 10 in order to impart information that only public authorities have access to.[138]

However, it remains unclear under which conditions this obligation exists, such as to which information it applies, what are the possible exceptions, and to whom information should be communicated.[139] More in detail, it has been argued that the ECtHR has recognised that a right to access to information for individual purposes may exist but in different provisions of ECHR. Namely, the ECtHR has most com-

---

[134] Santini (2004), p. 201. According to the Advocate General Léger in his Opinion delivered on 10 July 2001, *Council of the European Union v Heidi Hautala*, C-353/99, para. 55–59, "it is important to emphasise this convergence of national laws since to my mind it constitutes a decisive reason for recognizing the existence of a fundamental principle of a right of access to information held by Community institutions. (…) Those national rules, although the content of their corresponding legal systems are not necessarily the same, demonstrate a common conception in most of the Member States, which Advocate General Tesauro has described as follows: 'it is no longer true that everything is secret except what is expressly stated to be accessible, but precisely the converse'. In the light of that conception of relations between those who govern and those who are governed, on which there is almost unanimous consensus within the European Union, it appears natural to me to accept that there exists a principle of access to information held by the national public authorities and that that principle is such that it would engender an equivalent principle at Community level".

[135] See Chap. 2, para. 2.4.

[136] Österdahl (2016), p. 66.

[137] *Leander v Sweden*, App. No 9248/81, (ECtHR, 26 March 1987). See Österdahl (2016), p. 67; Buijze (2013), p. 107. For an examination of the cases in which the ECtHR has dealt with the potential right of access to documents, see Österdahl (2016) pp. 66 ff.

[138] Buijze (2013), p. 112; Österdahl (2016), p. 77.

[139] Buijze (2013), p. 112.

monly based a right of access in Article 8 ECHR, which provides *the right to respect for private and family life*.

In the *Gaskin* case,[140] the ECtHR ruled about the access of information "containing highly personal aspects of the applicant's childhood, development and history", and recognised a right to access certain personal information under Article 8 ECHR.[141] The Court specified that "although this right is not absolute, there should be at least a system in place" to balance the rights of individuals to personal information "against the public interest in keeping it secret".[142]

Furthermore, in the more famous *Guerra* case[143] the ECtHR has recognised the right to access information about the applicants' "environment". In this case, the applicants were looking for information about the environmental risk posed by a chemical factory close to their hometown (Manfredonia, Italy), and the emergency plans the company was required to adopt and submit to the municipality. In verifying whether the national authorities had taken "the necessary steps to ensure effective protection", the ECtHR observed that the applicants had not received the information that would have enabled them to assess the risks they could have run if they continued to live in their hometown.[144] The Court recognised the rights of the applicants to make informed decisions and the illegitimate authority's refusal to supply the information requested on the grounds of an emergency plan that was never submitted by the factory. The ECtHR reiterated its previous case law, pointing out that "that freedom to receive information (…) basically prohibits a government from restricting a person from receiving information that others wish or may be willing to impart to him. That freedom cannot be construed as imposing on a State, in circumstances such as those of the present case, positive obligations to collect and disseminate information of its own motion".[145] Accordingly, a failure to inform about a risk to an individual's health amounts to an infringement of the right to private life enshrined in Article 8 ECHR.[146]

Nonetheless, the European legislative framework rooted in the Treaties and the Charter of Fundamental Rights leads to the qualification of the right of access as a fundamental right. As noted before, Article 15 TFEU establishes this right which has been operationalised by the Regulation (EC) No 1049/2001.

In the *Interporc* case, the CJEU takes into account the legislative developments and it remarks that the Regulation (EC) No 1049/2001 lays down the conditions to enable citizens to participate more closely in the decision-making process and to guarantee that the administration "is (…) more accountable to the citizens in a

[140] *Gaskin v the United Kingdom*, App.No 10454/83 (ECtHR, 7 July 1989) Series A No. 160.

[141] Buijze (2013), p. 228; Hins and Voorhoof (2007), p. 119.

[142] Buijze (2013), p. 228. For a different interpretation of the Gaskin case, see O'Neill (1998), p. 421.

[143] *Guerra and Others v Italy* [GC] 14967/89 (ECtHR, 19 February 1998) 116/1996/735/932.

[144] Buijze (2013), p. 230; Hins and Voorhoof (2007), p. 119.

[145] *Guerra and Others v. Italy*, para. 53.

[146] See also Hins and Voorhoof (2007), p. 117; Santini (2004), p. 202; Kocharov (2009), p. 344.

democratic system and to contribute to strengthening the principle of democracy and respect of fundamental rights".[147]

Finally, the right of access "was unequivocally upgraded to a fundamental right"[148] with its incorporation in Article 42 Charter of Fundamental Rights of the European Union.

The coincidence between the right embedded in Article 15 TFEU and the one included in the Charter further leads scholars to consider the right of access as a fundamental right.[149]

---

[147] Judgment of 6 March 2006, *Interporc v Commission*, C-41/00 P, ECLI:EU:C:2003:125, para. 39. See Chap. 2, para. 2.2.2. For an interpretation of this case in the light of a general principle of access, see Lenaerts (2004), p. 321.

[148] Lenaerts (2004), p. 322. See also Santini (2004), p. 205; Labayle (2013), pp. 5 ff. Once again, it is worth to mention the Advocate General Léger's words in his Opinion delivered on 10 July 2001, *Council of the European Union v Heidi Hautala*, C-353/99, para. 80–86: "(…) The Charter of Fundamental Rights precludes it from being regarded as a mere list of purely moral principles without any consequences. It should be noted that those values have in common the fact of being unanimously shared by the Member States, which have chosen to make them more visible by placing them in a charter in order to increase their protection. The Charter has undeniably placed the rights which firm its subject matter at the highest level of values common to the Member States. (…) As the solemnity of its form and the procedure which led to its adoption would give one to assume, the Charter was intended to constitute a privileged instrument for identifying fundamental rights. It is a source of guidance as to the true nature of the Community rules of positive law. (…) The content of the right of access to documents, as set out in the Charter, is thus delimited by the provisions of Article 255 EC. That delimitation is the logical consequence of the difference in legislative value between the Charter and the provisions of the Treaty. That should not, however, cause us to overlook the fundamental nature of that right, as affirmed by the Member States of the Union at the time it was introduced into the Charter. Although not enshrining a positive right itself, Article 42 of the Charter confers on that right a quality which should provide guidance for its interpretations. I consider that where it is decided that a right should be classified as a fundamental right the authorities responsible for applying it are under a strict requirement to give it the wide interpretation demanded by its true nature".

[149] See Explanations relating to the Charter of Fundamental Rights, OJ C 303, 14.12.2007, pp. 17–35, where it is stated that Article 42 Charter of Fundamental Rights "has been taken over from Article 255 of the EC Treaty, on the basis of which Regulation (EC) No 1049/2001 has subsequently been adopted. The European Convention has extended this right to documents of institutions, bodies and agencies generally, regardless of their form (see Article 15(3) of the Treaty on the Functioning of the European Union). In accordance with Article 52(2) of the Charter, the right of access to documents is exercised under the conditions and within the limits for which provision is made in Article 15(3) of the Treaty on the Functioning of the European Union".

3.3.3 A Short Introduction on Regulation (EC) No 1049/2001 and Regulation (EC) No 1367/2006

Given the consideration of access to documents as a fundamental right of the EU law and the relevance of the legislative framework on access to documents for EU food law, we must describe some general aspects of Regulation (EC) No 1049/2001 (or "Access Regulation").

It is worth noting that Recital 2 of the Access Regulation's preamble states that openness "enables citizens to participate more closely in the decision-making process and guarantees that the administration enjoys greater legitimacy and is more effective and more accountable to the citizens in a democratic system" and it "contributes to strengthening the principles of democracy and the respect for fundamental rights" as laid down in Article 6 of the EU Treaty and in the Charter of Fundamental Rights.[150]

The Access Regulation governs the *regime* of access to documents in EU law. The Regulation follows a general pattern formed by defining provisions, the meaning of the term "document", the beneficiaries and the exceptions that limit access which is similar to other regimes on access to information.[151]

Article 2 Regulation (EC) No 1049/2001 gives a right to access to all documents of the European institutions in *all areas of EU activity*.

The Access Regulation also applies to *EU bodies and agencies*, according to the fact that agencies' internal regulations must comply with it. Pursuant to the Joint declaration relating to Regulation (EC) No 1049/2001 of the European Parliament and of the Council of 30 May 2001 regarding public access to European Parliament, Council and Commission documents, the three institutions agree that "the agencies and similar bodies created by the legislator should have rules on access to their documents which conform to those of this Regulation. To this effect, the European Parliament and the Council welcome the Commission's intention to propose, as soon as possible, amendments to the acts establishing the existing agencies and bodies and to include provisions in future proposals concerning the establishment of such agencies and bodies. They undertake to adopt the necessary acts rapidly".[152]

---

[150] As referred by Curtin and Leino-Sandberg (2016), p. 7, in its "landmark ruling in *Turco*, which clearly lays down the main principle of public access and its linkage with the principle of democracy" the Court expresses that "it is for the Council to balance the particular interest to be protected by non-disclosure of the document concerned against, inter alia, the public interest in the document being made accessible in the light of the advantages stemming (…) from increased openness, in that this enables citizens to participate more closely in the decision-making process and guarantees that the administration enjoys greater legitimacy and is more effective and more accountable to the citizen in a democratic system".

[151] Craig (2018), p. 394.

[152] Joint declaration relating to Regulation (EC) No 1049/2001 of the European Parliament and of the Council of 30 May 2001 regarding public access to European Parliament, Council and Commission documents, OJ L 145 of 31.5.2001, p. 43, it continues "2. The European Parliament, the Council and the Commission call on the institutions and bodies not covered by paragraph 1 to

In practice, the EU agencies apply the Regulation on access based on their own founding regulation. For instance, the recent Regulation (EU) 2019/1381 on the transparency and the sustainability of the EU risk assessment in the food chain provides that the right of access to documents as established in Regulation (EC) No 1049/2001 and in Regulation (EC) No 1367/2006 on access to environmental information[153] applies to documents held by the European Food Safety Authority (EFSA). Those regulations are of particular relevance for this book. In particular, pursuant to Article 41 Regulation 178/2002, EFSA's Management Board adopted the Decision (No. wp200327-a2) laying down *practical arrangements for implementing regulations on access to documents* repealing the Decision of 16 September 2003, further superseded in 2023 with the Decision of the Management Board laying down *practical arrangements for implementing Regulation (EC) No 1049/2001 and Articles 6 and 7 of Regulation (EC) No 1367/2006.*[154] Therefore, EFSA is expected to grant access to documents upon the applicant's request. Later in the book, we explore the implications of access to documents released by EFSA and the conflicting interests emerging from transparency and confidentiality rules.

Key case law about access to documents where third parties requested to access studies about the glyphosate approval or renewal process have significantly contributed to the debate about transparency. Particularly, they focused on the transparency of the authorisation procedure based on the scientific assessment and the challenge of balancing this with the disclosure of information included in the authorisation

adopt internal rules on public access to documents which take account of the principles and limits in this Regulation". For an overview on the various acts adopted and some issues posed by peculiarities concerning ECB and EBI see Santini (2004), pp. 200 ff. Neamtu and Dragos (2019), p. 17.

[153] Regulation (EC) No 1367/2006 of the European Parliament and of the Council of 6 September 2006 on the application of the provisions of the Aarhus Convention on Access to Information, Public Participation in Decision-making and Access to Justice in Environmental Matters to Community institutions and bodies, OJ L 264, 25.9.2006, p. 13–19.

[154] Decision of the Management Board laying down practical arrangements for implementing Regulation (EC) No 1049/2001 and Articles 6 and 7 of Regulation (EC) No 1367/2006 (Available at https://www.efsa.europa.eu/sites/default/files/documents/wp200327-a2.pdf). See Chap. 5. Before the Transparency Regulation, the application of Regulation 1049/2001 has been extended to EFSA by an amendment to Article 41 General Food Law (namely, Article 1(3) Regulation (EC) No 1642/2003 of the European Parliament and of the Council of 22 July 2003 amending Regulation (EC) No 178/2002 laying down the general principles and requirements of food law, establishing the European Food Safety Authority and laying down procedures in matters of food safety, OJ L 245, 29.9.2003, pp. 4–6). Following this amendment and in accordance with Article 38(3) and 39(4) General Food Law, the EFSA Management Board adopted a series of decisions for the EFSA's internal implementation of the transparency and confidentiality rules. Among them, we must recall: Decision concerning access to documents of 16.9.2003 (repealed); Code of good administrative behavior 16.9.2003-11 (available at www.efsa.europa.eu/sites/default/files/corporate_publications/files/admincode.pdf.), Openness, transparency and confidentiality 16.09.2003, 13 (available at https://www.efsa.europa.eu/sites/default/files/corporate_publications/files/transparencyprinciples.pdf.); Decision concerning implementing measures of transparency and confidentiality requirements of 10.3.2005 (repealed). See also Kocharov (2009), p. 341.

dossiers.[155] The reform of the transparency rules adopted with Regulation (EU) 2019/1381 analysed also address this context.[156]

The right of access applies to "existing documents in possession of the institutions". It embraces "documents drawn up by those institutions" and documents "received from third parties, including the other EU institutions" as well as Member States.[157] As regards the scope, it applies to the "*information* contained in documents rather than to the documents as such".[158] The CJEU also provided that *partial access* must be possible, such as by editing parts of the documents.[159] Consequently, Article 4 (6) Regulation (EC) No 1049/2001 takes into account this principle.

As regards the procedure, access to documents requires a written application that must be processed according to Articles 6 and 7 Regulation (EC) No 1049/2001. In case of refusal, it is possible to ask the institution to reconsider. The case can be eventually brought before the General Court or the European Ombudsman as Article 8 Regulation (EC) No 1049/2001 foresees. Article 9 deals with sensitive documents, providing that they should be classified as secret in accordance with the institution concerned.[160]

The most discussed aspect among scholars consists of the exceptions' regime. The exceptions in Article 4 Regulation (EC) No 1049/2001 derogate from the principle of the "widest possible public access to documents" and must therefore be *interpreted narrowly*, as stated by the European Court of Justice.[161] Moreover, the

---

[155] Röttger-Wirtz (2020), p. 26, Vanuzzo (2024), p. 189. Of particular importance is the topic of access requests for the studies of the applicant see Chap. 5.

[156] See Chap. 4.

[157] Klamert (2019b), p. 401.

[158] Klamert (2019b), p. 401. See Judgment of the Court of First Instance (First Chamber) of 19 July 1999, *Heidi Hautala v Council of the European Union*, T-14/98, para. 23. Craig (2018), p. 394; Curtin (2000), pp. 16–17.

[159] The principle was firstly recognised in Judgment of the Court of First Instance (First Chamber) of 19 July 1999, *Heidi Hautala v Council of the European Union*, T-14/98, para. 23. Craig (2018), p. 394; Klamert (2019b), p. 401. The CFI adopted the same approach in Judgment of the Court of First Instance (First Chamber, extended composition) of 13 April 2005, *Verein für Konsumenteninformation v Commission of the European Communities*, Case T-2/03, ECLI:EU:T:2005:125, where a request pertained a large number of documents or a large file, and the CFI held that to safeguard the good administration's interest, the institution must retain the right to balance the interest in access and the burden of work. According to Santini (2004), p. 215, the Regulation (EC) No 1049/2001 seems to recall the case law on the partial access even in the notion of documents as defined in Article 3 (a) Regulation 1049/2001, which foresees that "'document shall mean any content whatever its medium (written on paper or stored in electronic form or as a sound, visual or audiovisual recording) concerning a matter relating to the policies, activities and decisions falling within the institution's sphere responsibility".

[160] Sensitive documents according to Article 9 Regulation (EC) No 1049/2001 should be classified as "TRÈS SECRET/TOP SECRET", "SECRET", or "CONFIDENTIEL". See Klamert (2019b), p. 401; Santini (2004), pp. 221–224.

[161] Judgment of the Court (Grand Chamber) of 1 July 2008, *Kingdom of Sweden and Maurizio Turco v Council of the European Union*, Joined cases C-39/05 P and C-52/05 P, ECLI:EU:C:2008:374, para. 36 and Judgment of the Court (Grand Chamber) of 18 December 2007, *Kingdom of Sweden*

fact that a case falls within an exception does not justify a refusal. Thus, the likelihood of the interest protected by the exception must be "reasonably foreseeable and not purely hypothetical".[162]

Therefore, the "most pressing characteristic"[163] of access is the many limitations to which it is subject arising from other public or private interests.

According to Article 4 (1) Access Regulation, the institution shall refuse access when the release of a document would undermine the protection of (a) the *public interest*, namely public security, defence and military matters, international relations and the financial, monetary, or economic policy of the Union or a Member State and (b) *privacy and integrity of individuals*, particularly in accordance with the EU legislation on the protection of personal data. These exceptions are mandatory.[164]

By contrast, Article 4 (2) Access Regulation foresees that access to a document shall be refused when disclosure undermines *commercial interests* of a natural or legal person, including *intellectual property*, court proceedings and legal advice, and the purpose of inspections, investigations and audits.[165]

Furthermore, Article 4 (3) Access Regulation provides the "*space to think*" exception, namely that access to a document drawn up by an institution for internal use or received by an institution referred to matters where a decision has not been taken by the institutions, shall be refused if the document's disclosure would seriously undermine the institution's decision-making process.[166]

In all of these circumstances, the regulation allows access even in case the document relates to a protected interest whether there is an *overriding public interest in disclosure*. Therefore, the exceptions laid down in Article 4 (2) and (3) are not mandatory as the one in the previous paragraph of the same article, but they require a *balancing of interests*.[167] As will be seen further in this book, the exception related

v Commission of the European Communities and Others, Case C-64/05 P, ECLI:EU:C:2007:802, para. 66. See Klamert (2019b), p. 401.

[162] Judgment of the Court (Grand Chamber) of 1 July 2008, *Kingdom of Sweden and Maurizio Turco v Council of the European Union*, para 43. See Craig (2018), p. 396; Klamert (2019b), p. 401.

[163] Abazi and Tauschinsky (2015), p. 87.

[164] A greater number of case law is about these exceptions. We recall the case law referred by Craig (2018), p. 396; Santini (2004), pp. 225–240; Abazi and Hillebrandt (2015), pp. 825–845; Neamtu and Dragos (2019), pp. 37–42. As regard the privacy of individuals, of particular importance is the Judgment of the Court (Second Chamber) of 16 July 2015, *ClientEarth and Pesticide Action Network Europe (PAN Europe) v European Food Safety Authority*, Case C-615/13, ECLI:EU:C:2015:489. See Oster (2016), pp. 272–277; González Vaqué (2015), pp. 1–5; Giulimondi (2015), p. 177. See also Chap. 5, para. 5.2.1.1.

[165] On copyright issues, Article 16 Access Regulation provides that the Regulation itself "shall be without prejudice to any existing rules on copyright which may limit a third party's right to reproduce or exploit released documents". Even if it can "inform institutional choices", this is not provided as an exception to disclosure. See Korkea-Aho and Leino (2017), p. 1069.

[166] Klamert (2019b), p. 402.

[167] Klamert (2019b), p. 402. The judicial approach to the balancing test was settled in the Judgment of the Court (Grand Chamber) of 1 July 2008, *Kingdom of Sweden and Maurizio Turco v Council of the European Union*, See Chap. 5, para. 5.2.1.1.

to the protection of commercial interests is of particular relevance with regard to transparency in the food law domain.[168]

As regards third-party documents, Articles 4 (4) Access Regulation only requires that the third-party must be consulted before disclosure. By contrast, pursuant to Article 4 (5) Access Regulation a Member State can demand that documents originating from it only be disclosed after its prior agreement. As clarified by the CJEU in the cases *Sweden v Commission* and *IFAW*, it does not confer a general right of veto to a Member State, but it gives the right to expose whether a document may fall under an exception.[169]

Leaving aside an in-depth analysis of the Regulation (EC) No 1049/2001,[170] we shall, however, recall that in accordance with Article 17(1), each institution shall publish annually a report for the preceding year, including "the number of cases in which the institution refused to grant access to documents, the reasons for such refusals and the number of sensitive documents not recorded in the register".[171]

The regulatory framework of access to documents does not only encompass the Access Regulation cornerstone. Therefore, as regards *access to information related to the environment*, the Aarhus Convention comes into play. More in detail, the European Union has established a body of rules enshrining the principle that the public has a right of access to environmental information by adopting on 25 June 1998 the United Nations Economic Commission for Europe (UNECE)'s Convention on access to information, public participation in decision-making and access to justice in the environmental matters ("Aarhus Convention").[172] That legislation added to the relevant provisions of Regulation (EC) No 1049/2001, and to the relevant provisions regarding access to information on sector specific EU legislation adopted in relation to environmental protection.[173] The Aarhus Convention articulates a right-based framework focused on the procedural rights of citizens, with access to information in support of access to decision making and access to justice in environmental matters.[174] Therefore, access to information represents an essential prerequi-

---

[168] See Chap. 5, para. 5.2.1.2.

[169] See Judgment of the Court (Grand Chamber) of 18 December 2007, *Kingdom of Sweden v Commission of the European Communities and Others*, Case C-64/05 P. and Judgment of the Court (Third Chamber), 21 June 2012, *IFAW Internationaler Tierschutz-Fonds gGmbH v European Commission*, Case C-135/11 P, ECLI:EU:C:2012:376, and Klamert (2019b), p. 402.

[170] See, for the practical dimensions, Labayle (2013), pp. 33–37; Curtin and Leino-Sandberg (2016), p. 32; Neamtu and Dragos (2019), pp. 11 ff.

[171] See, for instance, the Report from the Commission on the application in 2023 of Regulation (EC) No 1049/2001 regarding public access to European Parliament, Council and Commission documents, COM/2024/266 final.

[172] Convention on Access to Information, Public Participation in Decision-Making and Access to Justice in Environmental Matters. On the adoption's context, see Mason (2014), pp. 83–105.

[173] As reported also by CJEU Research and Documentation Directorate (2017), the CJEU has developed a substantial body of case-law in various judicial proceedings. See also Fulop (2024).

[174] Mason (2014), p. 88. The right of access to information represents the so-called first pillar of the Aarhus Convention. The second pillar regulates public participation in decision-making process

site for other environmental rights. In the EU legal framework, the abovementioned Regulation (EC) No 1367/2006 ("Aarhus Regulation") and Directive 2003/4/EC[175] implement the Aarhus Convention by guaranteeing the right of access to environmental information received or produced by the institutions or bodies, and by setting out the terms and conditions to exercise that right.[176] As mentioned, the Aarhus Regulation extends the application of the Regulation (EC) No 1049/2001 to "any request by an applicant for *access to environmental information* held by Community institutions and bodies without discrimination as to citizenship, nationality or domicile and, in the case of a legal person, without discrimination as to where it has its registered seat or an effective centre of its activities".[177] Therefore, it is crucial to understand the meaning of access to environmental information and the possible exceptions to it laid down by the legislation to apply it in practical cases.[178] In short, the legal framework on access to information held by the EU authorities is a "dual track one" composed of the Access Regulation as amended and supplemented by the Aarhus Regulation.[179]

3.4 Brief Concluding Remarks on Transparency as a Fundamental Principle in EU Law

Based on the aforementioned arguments drawn on legal doctrine and EU jurisprudence, we must deduce that in the EU there is not a general principle of transparency encompassing all the possible features that transparency could have. Transparency has emerged as a principle across different areas of EU law. However, the debate on establishing a general principle of transparency has mainly focused on "passive transparency" especially after the changes in the legislative framework on access to documents mentioned earlier.[180]

---

that could have an impact on the environment, while the third pillar concerns access to justice in environmental matters. See Garcon (2013), pp. 78–90. On access to justice, see more recently Colombo (2024) and Eliantonio and Richelle (2024).

[175] Directive 2003/4/EC of the European Parliament and of the Council of 28 January 2003 on public access to environmental information and repealing Council Directive 90/313/EEC. Official Journal L 041, 14/02/2003 P. 0026 – 0032.

[176] See Article 1 Regulation (EC) No 1367/2006. In October 2021, the EU adopted Regulation (EU) 2021/1767 that amended Regulation 1367/2006 to allow for better public scrutiny of EU acts affecting the environment by NGOs and other members of the public.

[177] Article 3 Regulation (EC) No 1367/2006. Emphasis added.

[178] About the CJEU's recent case law in the context of access and confidentiality in Chap. 5, para. 5.2.1.1.

[179] This "dual-track solution" was "deemed necessary at the Community level" due to the lack of full compliance of the Access Regulation's regime with the firs pillar of the Aarhus Convention. See de Abreu (2008), p. 187. See also Wolf (2013), pp. 475–449.

[180] See para 3.1 and 3.2.

It could be concluded that two main aspects appear to arise in the effort to better understand transparency in the EU law. Namely, the principle of transparency is also inherent in well-established legal principles and can aid in the interpretation of those.[181] Furthermore, among scholars, there is a rooted perception that only access to documents qualifies as "candidate" for recognition as a general principle of EU law.[182]

References

Abazi V, Hillebrandt M (2015) The legal limits to confidential negotiations: recent case law developments in council transparency: access info Europe and in 't veld. Common Mark Law Rev 52(3):825–846

Abazi V, Tauschinsky E (2015) Reasons of control and trust: grounding the public need for transparency in the European Union. Utrecht Law Rev 11:78–90

de Abreu S (2008) Passive access to environmental information in the EU—an analysis of recent legal developments. Eur Energy Environ Law Rev 17(4):186–198

Adamski D (2012) Approximating a workable compromise on access to official documents: the 2011 developments in the European courts. Common Mark Law Rev 49:521–558

Addink GH (2009) The transparency principle in the framework of the WTO. Indonesian J Int Law 6 (2):232–243

Alemanno A (2014) Unpacking the principle of openness in EU law. Transparency, participation and democracy. Eur Law Rev 39(1):72–90

Blanke HJ (2013) Article 1. Establishment and functioning of the Union. In: Blanke HJ, Mangiameli S (eds) The treaty on European Union (TEU): a commentary. Springer, Berlin, p 73. https://doi.org/10.1007/978-3-642-31706-4_2

Bovis CH (2009) The effects of the principles of transparency and accountability on public procurement regulation. In: Hofmann H, Türk AH (eds) Legal challenges in EU administrative law. Towards an integrated administration. Edward Elgar Publishing, Cheltenham, UK, pp 288–321. https://doi.org/10.4337/9781848449206

Buijze A (2013) The principle of transparency in EU law. Dissertation, Utrecht University

CJEU Research and Documentation Directorate (2017) Fact Sheet on Public Access to Environmental Information. Available via https://curia.europa.eu/jcms/upload/docs/application/pdf/2018-05/fiche_thematique_-_environnement_-_en.pdf. Accessed 7 June 2025

Colombo E (2024) Principles of EU law in climate litigation. China-EU Law J. https://doi.org/10.1007/s12689-024-00108-9

Craig P (2018) EU administrative law, 3rd edn. Oxford University Press, Oxford

Craig P, De Búrca G (2020) EU law: text, cases, and materials, 7th edn. Oxford University Press, Oxford

Curtin D (2000) Citizens' fundamental right of access to EU information: an evolving digital Passe-partout. Common Mark Law Rev 37:7–41

Curtin D, Leino-Sandberg P (2016) Openness, Transparency and the Right of Access to Documents in the EU. In-Depth Analysis. Directorate General for Internal Policies. Policy Department C: Citizen's rights and Constitutional Affairs. Petitions. Available via https://www.europarl.europa.eu/RegData/etudes/IDAN/2016/556973/IPOL_IDA(2016)556973_EN.pdf. Accessed 23 Dec 2024

[181] See para. 3.2.

[182] See para. 3.3.2.

De Witte B (2000) Institutional principles: a special category of general principles of EC law. In: Bernitz U, Nergelius J (eds) General principles of European community law. Kluwer Academic Publishers, Dordrecht, pp 143–162

Diamandouros PN (2006) Transparency, accountability, and democracy in the EU, lecture by the European ombudsman, Professor P. Nikiforos Diamandouros, at the School of Advanced International Studies of the Johns Hopkins University, Bologna, Italy, 17 October 2006. Available at https://www.ombudsman.europa.eu/en/speech/en/349. Accessed 7 June 2025

Eliantonio M, Richelle J (2024) Access to justice in environmental matters in the EU legal order: the "Sectoral" turn in legislation and its pitfalls. Eur Pap 9(1):261–274

European Council (2019) A new strategic agenda 2019–2024. Available via https://www.consilium.europa.eu/media/39914/a-new-strategic-agenda-2019-2024.pdf. Accessed 7 June 2025

Fulop S (2024) Case law of the Court of Justice of the European Union on access to environmental information. Legal analysis Justice and Environment 2024. Available via https://justiceandenvironment.org/wp-content/uploads/2024/11/CJEU-ATOI-Case-Law_final-for-web.pdf. Accessed 18 Apr 2025

Garcon G (2013) The rights of access to justice in environmental matters in the EU - the third pillar of the Aarhus convention. Eur Food Feed Law Rev 8(2):78–90

Giulimondi F (2015) Corte di Giustizia dell'Unione europea (Sezione seconda), sentenza 16 luglio 2015, causa C-615/13 P, ClientEarth e Pesticide Action Network Europe (PAN Europe) c. European Food Safety Authority (EFSA). Diritto di accesso e trasparenza: veri baluardi della credibilità delle Istituzioni europee. DPCE Online 24(4):171–180

González Vaqué L (2015) European Court of Justice rules in favour of greater transparency in accessing EFSA data. Available at https://scispace.com/pdf/european-court-of-justice-rules-in-favour-of-greater-aphvj279bv.pdf. Accessed 18 Apr 2025

Groussot X, Hettne J, Petursson GT (2017) General principles and the many faces of coherence: between law and ideology in the European Union. In: Vogenauer S, Weatherill S (eds) General principles of law. European and comparative perspectives. Hart Publishing Ltd, Oxford, pp 77–103

Harden I (2009) The revision of regulation 1049/2001 on public access to documents. Eur Public Law 15:239–256

Harlow C (1999) Freedom of information and transparency as administrative and constitutional rights. Cambridge Yearb Eur Leg Stud 2:285–302. https://doi.org/10.5235/152888712802815860

Heliskoski J, Leino P (2006) Darkness at the break of noon: the case law on regulation no. 1049/2001 on access to documents. Common Mark Law Rev 43:735–781

Hillebrandt MZ, Curtin D, Meijer A (2014) Transparency in the EU Council of ministers: an institutional analysis. Eur Law J 20:1–20

Hins W, Voorhoof D (2007) Access to state-held information as a fundamental right under the European convention on human rights. Eur Const Law Rev 3:114–126

Hoffman H, Leino-Sandberg P (2019) An Agenda for transparency in the EU. European Law Blog, October 23, 2019. https://europeanlawblog.eu/2019/10/23/an-agenda-for-transparency-in-the-eu/. Accessed 18 Apr 2025

Karageorgou V (2012) Transparency Principle as an Evolving Principle of EU Law: Regulative Contours and Implications 1-18 available at https://www.researchgate.net/publication/350671145_Transparency_principle_as_an_evolving_principle_of_EU_law_Regulative_contours_and_implications. Accessed 27 Nov 2024

Klamert M (2014) The principle of loyalty in EU law. Oxford University Press, Oxford

Klamert M (2019a) Article 1 TEU. In: Kellerbauer M, Klamert M, Tomkin J (eds) The EU treaties and the charter of fundamental rights: a commentary. Oxford University Press, Oxford, pp 7–25

Klamert M (2019b) Article 15 TFEU. In: Kellerbauer M, Klamert M, Tomkin J (eds) The EU treaties and the charter of fundamental rights: a commentary. Oxford University Press, Oxford, pp 397–404

Kocharov A (2009) Data ownership and access rights in the European food safety authority. Eur Food Feed Law Rev 4 (5):335–346

Korkea-Aho E, Leino P (2017) Who owns the information held by EU agencies: weed killers, commercially sensitive information and transparent and participatory governance. Common Mark Law Rev 54(4):1059–1092

Labayle H (2013) Openness, transparency and the right of access to documents and information in the European Union. Directorate-General for Internal Policies. Citizens' Rights and Constitutional Affairs (PE 493/035). Available via https://www.europarl.europa.eu/studies/. Accessed 3 Dec 2024

Leino P (2002) Case C-353/99 P, Council v. Heidi Hautala, Judgment of the Full Court of 6 December 2001, Nyr, Appeal Against the Judgment of the Court of First Instance of 19 July 1999 in Case T-14/98. Common Mark Law Rev 39:621–632

Lenaerts K (2004) 'In the Union We Trust': trust-enhancing principles of Community law. Common Mark Law Rev 41:317–343

Lock T (2019a) Article 41 CFR. In: Kellerbauer M, Klamert M, Tomkin J (eds) The EU treaties and the charter of fundamental rights: a commentary. Oxford University Press, Oxford p, pp 2204–2207

Lock T (2019b) Article 42 CFR. In: Kellerbauer M, Klamert M, Tomkin J (eds) The EU treaties and the charter of fundamental rights: a commentary. Oxford University Press, Oxford, p 2208

Lodge J (2003) Transparency and EU Governance: Balancing Openness with Security, Journal of Contemporary European Studies, 11 (1), 95-117

Mason M (2014) So far but no further? Transparency and disclosure in the Aarhus convention. In: Gupta A, Mason M (eds) Transparency in global environmental governance. Critical perspectives. MIT Press, Cambridge, pp 83–106

Neamtu B, Dragos DC (2019) Freedom of Information in the European Union: Legal Challenges and Practices of EU Institutions. In: Dragos DC, Kovač P, Marseille AT (eds) The Laws of Transparency in Action. A European Perspective. Palgrave Macmillan, Cham, pp 11–70

Neuvonen PJ, Ziegler KS (2022) General principles in the EU legal order: past, present and future directions. In: Ziegler KS, Neuvonen PJ, Moreno-Lax V (eds) Research handbook on general principles in EU law, pp 7–24

O'Neill M (1998) The right of access to community-held documentation as a general principle of EC law. Eur Public Law 4:403–432

Oster J (2016) Of Money, Poison and Secrets: Balancing Freedom of Information against Data Protection. European Data Protection Law Review 2(2):272–277

Österdahl I (2016) Transparency as part of a European rule of law. In: Schroeder W (ed) Strengthening the rule of law in Europe: from a common concept to mechanisms of implementation. Hart Publishing Ltd, Oxford, pp 61–79

Prechal S, de Leeuw M (2007) Dimensions of transparency: the building blocks for a new legal principle? Rev Eur Adm Law 0:51–62. https://doi.org/10.7590/REAL_2007_00_03

Röttger-Wirtz S (2020) D.2.3 Glyphosate case study: Contribution to the REconciling sCience, Innovation and Precaution through the Engagement of Stakeholders Project (RECIPES). Horizon 2020 Framework Programme of the European Union under Grant Agreement no. 824665. Available via Maastricht University. https://recipes-project.eu/sites/default/files/2020-10/_CS6_Glyphosate.pdf. Accessed 23 Dec 2024

Rovetta D, Gambardella M (2014) Access to EU antidumping/subsidy confidential files: the EU law 'Zwartveld Sincere Cooperation Doctrine', surprise what you are looking for already exists. Glob Trade Cust J 9:338–340

Sabbatini M (2003) Commento agli articoli 38, 39, 40, 41, 42 del reg. (CE) n. 178/2002. La sicurezza alimentare nell'Unione Europea. Commentario a cura dell'IDAIC al reg. (CE) n. 178/2002. Le nuove leggi civili commentate 1–2:383–405

Santini A (2004) Il principio di trasparenza nell'ordinamento dell'Unione Europea. Giuffrè, Milano

Semmelmann C (2013) General principles in EU law between a compensatory role and an intrinsic value. Eur Law J 19:457–487

Tomkins A (1999) Transparency and the emergence of a European administrative law. Yearbook of European Law, Oxford 19(1):217–256

Tridimas T (2006) The general principles of EU law, 2nd edn. Oxford University Press
Goudappel F, Van den Brink A (2011) Processes of transnationalization of administrative values: administrative regulation and transparency in the EU. In: Anthony G, Auby JB, Morison J, Zwart T (eds) Values in global administrative law. Hart Publishing, Oxford and Portland, pp 373–392
Van der Meulen B (2020) Authorisation requirements. In: Van der Meulen B, Wernaart B (eds) EU food law handbook, European Institute for Food law Series, vol 13. Wageningen Academic Publishers, Wageningen, 205–244. https://doi.org/10.3920/978-90-8686-903-9
Vandamme T (2006) Joined Cases C-154/04 and C-155/04, Alliance for National Health and Nutri-Link v. Secretary of State for Health and the National Association of Health Stores Health Food Manufacture. Leg Iss Econ Integr 33(3):305–318
Vanuzzo M (2024) Public health competencies and transparency in the internal market: a focus on the marketing authorisations in the EU pharmaceutical reform. Federalismi.it 2:176–190
Vesterdorf B (1998) Transparency - not just a vogue word. Fordham Int Law J 22:902–929
Volpato A, Eliantonio M, Wright K (2023) Transparency and participation in the face of scientific uncertainty: concluding remarks. Eur J Risk Regul 14(2):371–381
Von Bogdandy A (2010) Founding principles of EU law: a theoretical and doctrinal sketch. Revus 12:35–56
Vos E, Volpato A, Bellenghi G (2023) Independence and transparency policies of the European Food Safety Authority (EFSA), publication for the Committee on Environment, Public Health and Food Safety (ENVI), Policy Department for Economic, Scientific and Quality of Life Policies, European Parliament, Luxembourg. Available at https://www.europarl.europa.eu/RegData/etudes/STUD/2023/740080/IPOL_STU(2023)740080_EN.pdf. Accessed 7 May 2025
Weiss F, Steiner S (2007) Transparency as an element of good governance in the practice of the EU and the WTO. Fordham Int Law J 30:1545–1586
Wolf S (2013) Access to EU environmental information: EU compliance with Aarhus convention. ERA-Forum 14(4):475–449

Chapter 4
How Is the Principle of Transparency Enshrined in Current European Food Law?

Abstract In the first part, this chapter proposes a commented analysis of the General Food Law ("GFL"), moving from EU law to the specific EU food law domain. It further considers the policy issues behind its adoption to facilitate a wider understanding of the EU food law legal framework. Furthermore, after exposing the GFL structure and the "Fitness Check", this section focuses on the amendments introduced by Regulation (EU) 2019/1381 ("Transparency Regulation"). There, the analysis deals with the description of transparency obligations and the application of the "dimensions" framework shaped in Chap. 1. In the second part, this chapter investigates the judicial approach to transparency in the food sector. It examines how the CJEU Court of Justice of the European Union addresses transparency in its case law related to food (research question) by using software that supports qualitative content analysis (CAQDAS). Following an in-depth CJEU case law analysis, the chapter draws some conclusions. (This chapter contains also excerpts published in Vanuzzo (2021).).

4.1 Legislative Analysis: From the Internal Market to the New Transparency Rules: Introduction

The current European food law has been developed after a process started with the creation of the internal market. From the origin of the European Union, market integration was the primary objective, and food safety was touched upon only indirectly.[1] The main scope of the first regulations in food law was facilitating the free movement of goods through the internal market.[2]

[1] Vos (2013), p. 112.

[2] See De Boer (2019a), p. 247; Szajkowska (2009), p. 529. Articles 14 and 28–37 of the Treaty of functioning of the European Union (TFEU) regulate the free movements of goods. On the internal market, see Bestagno (2020), p. 1 ff.; Purnhagen (2014), pp. 315–342; Weatherill (2014), pp. 401–418. See also Commission Notice Guide on Articles 34–36 of the Treaty on the Functioning of the European Union (TFEU), Brussels 5.3.2021 C (2021) 1457 final.

M. Vanuzzo, *Transparency in EU Law*, Studies in European Economic Law and Regulation 29, https://doi.org/10.1007/978-3-032-02345-2_4

In the 1970s, the European Union sought to replace national legislation with a comprehensive EU-level regulations in the food sector, to harmonise it through vertical directives.[3] For example, the Directive related to cocoa and chocolate products intended for human consumption set how much cocoa a food product should contain.[4] In those years, the EU legislature adopted several rules about ingredients prohibited or permitted, taking into account technological advancements in some substances, such as food additives and flavourings.[5]

The European Union abandoned the vertical harmonisation strategy with the establishment of the principle of mutual recognition. In the landmarking *Cassis de Dijon* case, the CJEU advanced a foundational principle of internal market law, according to which products that have been legally produced and marketed in a Member State cannot be excluded from other Member States on the grounds of non-compliance with national regulations.[6] As a consequence, Member States are obliged to accept in their market food lawfully produced in another Member State and permit the circulation of goods, even in the absence of harmonisation.

During this second stage of the market-oriented phase of development, food law required harmonisation through horizontal directives.[7] This led to a change in the perception of the benefits of harmonisation: for Member States with stringent national standards, European legislation became "the best hope for raising neighbours' standards" to be able to achieve a level playing field without harming consumers protection.[8]

It has been argued that among the regulations adopted in those years, the one on HACCP included some transparency elements.[9] More in detail, the Council

---

[3] Vos (2013), p. 113; Van der Meulen (2013), p. 73; Van der Meulen (2020).

[4] Directive 73/241/EEC meanwhile replaced by Directive 2000/36/EC. See Van der Meulen (2020), p. 139.

[5] De Boer (2019a), p. 247.

[6] Judgment of the Court of 20 February 1979, *Rewe-Zentral AG v Bundesmonopolverwaltung für Branntweinudgment*, C-120/78, ECLI:EU:C:1979:42. See Van der Meulen (2020), p. 141. The case concerns a German chain of supermarket (Rewe) seeking to import the fruit liqueur Cassis de Dijon from France. This intention was opposed by the German authorities, who refused to authorise the import because the alcohol content was lower than the one pursuant to the German national standards (according to which a liqueur should contain at least 25% alcohol, while the Cassis contained only 20% alcohol). The German authorities, even though they knew it was a restriction on trade, justified the refusal arguing that alcoholic beverages with low alcohol content pose several risks. For instance, alcoholic beverages with low alcohol content could provoke a quicker tolerance for alcohol in the consumers. Moreover, since the taxes on alcohol are high, low-content alcohol beverages would benefit from an unfair competitive advantage and be saleable at lower prices than beverages produced in Germany. The Court found that Germany's authorities arguments did not meet the urgent need standard (rule of reason) requested for the exceptions of the Treaties. The Court concluded that in this case were not present valid justifications for restriction on the trade between Member States. See also Regan (2010), pp. 465–473; Purnhagen (2014), p. 315 ff.

[7] Van der Meulen (2013), p. 73.

[8] Van der Meulen (2020), p. 143.

[9] Albisinni (2015), p. 104.

Directive 93/43/EEC[10] gave legal relevance to internal control and external communication by introducing the principle of analysis and control over critical points.[11] Moreover, guidelines and voluntary handbooks establishing hygiene practices facilitated the distinctions among product techniques and products and prioritised the food businesses' responsibilities.[12] Since the HACCP method was based on the principle "analyse what you do, declare and document it", food producers were called to be transparent in their activities.[13]

A series of crises in the 1990s caused a collapse of consumers' trust in authorities, industry and science and triggered the end of the 'market oriented' phase.[14] As is known, the BSE crisis, despite it was not the worst,[15] caused a significant upheaval in the EU legal framework. Subsequent food scares, scandals, and animal diseases have accelerated the urgency to tackle food safety issues.[16] After a scientific advisory body to the British Government concluded that there might have been a link between BSE and the new variant of Creutzfeldt-Jacob disease, the Commission adopted a ban prohibiting the UK from exporting beef.[17] The UK initially responded with a non-cooperation policy with the EU Institutions. The European Parliament set up in July 1996 the Temporary Committee of Inquiry into BSE. The latter found that the European Union's internal management and decision-making procedures suffered severe deficiencies.[18] Furthermore, the relationship between scientific and political decisions was not transparent, partially due to the inefficiencies of the com-

---

[10] Council Directive 93/43/EEC of 14 June 1993 on the hygiene of foodstuffs, OJ L 175, 19.7.1993, now repealed by Regulation (EC) No 852/2004 of the European Parliament and of the Council of 29 April 2004 on the hygiene of foodstuffs, OJ L 139, 30.4.2004. On the HACCP, see Spirovska Vaskoska et al. (2020), pp. 304 ff.

[11] Albisinni (2015), p. 104.

[12] Albisinni (2015), p. 104.

[13] Albisinni (2015), p. 104. The Author points out that in that context "it is no more sufficient that the single food product is complying with the single prescription regarding that sort of food: all the food business structure and activity must be transparent, must be a *glass house*, where official control can look through".

[14] Van der Meulen (2013), p. 75.

[15] Van der Meulen (2013), p. 76. The author points out that in terms of death rate the toxic oil syndrome epidemic (TOS) occurred in Spain in the spring of 1981 "caused approximately 20,000 cases of a new illness". The poisoning was caused by fraud based on mixing consumption oil with vehicle oil.

[16] Van der Meulen (2013), p. 76. The author recalls several examples. An example of food safety scare is the Belgian dioxin crises caused by industry oil into animal feed and consequently in the food chain. Another example is when medroxyprogesterone acetate (MPA), a hormone used in contraceptive pills, entered in the food chain through pig feed to which was added sugar discharges from MPA production. A scandal example is the melamine crisis. As outbreak of animal disease, one can recall the foot-and-mouth disease, SARS, pig plague and avian influenza.

[17] Vos (2013), p.114. On the consequences for the internal market and the concerns expressed by the French Food Standard Agency (AFSSA), see Purnhagen (2015), p. 918.

[18] Vos (2013), p. 114. The Committee chaired by Manuel Medina Ortega was institute to investigate the actions of the agencies involved in the BSE crisis. The outcome is also called "Medina Ortega" report. See Van der Meulen (2013), p. 76.

mittee model in decision-making procedures.[19] It is interesting to note that the Report of the Temporary Committee of Inquiry into BSE recognised a significant *lack of transparency* in the BSE affair. It emphasised that "the rules governing the work of the scientific committees which advise the Commission should be reformed in such a way as to professionalise their organisation" and that "there should be improved guarantees concerning their independence".[20] In this context, the Commission also faced criticism for prioritising the interests of the food industry.

The Report provided concrete recommendations to improve the EU food law's structure, to which the Commission reacted by implementing the Committee's recommendations. First, the Commission published a Green Paper in 1997 on the general principles of food law in the European Union, which served as the first step in developing new legislation on food.[21] The Commission no longer aligned its food policy to the market integration but adopted a more coordinated strategy towards food safety. In January 2000, the Commission published its White Paper on Food Safety.[22] This document aimed to sustain consumers' trust by reviewing food legislation and establishing a comprehensive approach to food safety.[23]

In addition, the White Paper proposed the creation of a European Food Safety scientific body as a reference for the new regulatory framework. In the White Paper on Food Safety there is a clear intimation to transparency, as it emphasises that "*greater transparency* at all levels of Food Safety policy is the *golden thread* through the whole White Paper and it will contribute fundamentally to enhancing consumer confidence in EU Food Safety policy".[24]

---

[19] Vos (2013), p. 114.

[20] See Report of the Temporary Committee of Inquiry into BSE, set up by the Parliament in July 1996, on the alleged contraventions or maladministration in the implementation of Community law in relation to BSE, without prejudice to the jurisdiction of the Community and the national courts of 7 February 1997, A4-0020/97/A, PE 220.544/fin/A.

[21] The General Principles of Food Law in the European Union, Commission Green Paper, COM 97(176) final, Brussels 30.04.1997. See de Boer (2019a), p. 248. See also Albisinni (2015), p. 105 ff. The author recalls that in the same year the Council adopted the Regulation No 820/97 establishing a system for the identification and registration of bovine animals regarding the labelling of beef products. This Regulation mentioned in Recital 1 transparency "as a to be reached and as a tool to guarantee the stabilisation of markets".

[22] White Paper on Food Safety, COM (1999) 719 final, Brussels, 12 January 2000.

[23] Van der Meulen (2013), p. 77. The Annex to the White Paper is the Action Plan on Food Safety, "a list of 84 legislative steps that the Commission deemed necessary to create a regulatory framework capable of ensuring a high level of consumer's protection". Most of the 84 steps were subsequently adopted in the General Food Law.

[24] White Paper on Food Safety, COM (1999) 719 final, Brussels, 12 January 2000, p. 36. Emphasis added.

4.2 The General Food Law

The first step for the realisation of the White Paper on Food Safety's objectives has been the adoption of the Regulation (EC) No 178/2002 of the European Parliament and of the Council of 28 January 2002 which lays down the general principles and requirements of food law, establishes the European Food Safety Authority and lays down procedures in matters of food safety. As mentioned in the previous Chapters, this Regulation is called "General Food Law Regulation", or by the acronym "GFL". The General Food Law Regulation forms the basis of food and feed legislation, establishing a comprehensive framework.[25] To this end, it aims to accomplish the objectives of laying down the *principles*, *requirements* and *procedures* that underpin decision making on food and feed safety, covering all stages of food and feed production and distribution.

As Recital (4) and (5) explained, the differences between the Member States' food laws concerning concepts, principles and procedures may impede the free movement of food. Accordingly, it has been necessary to approximate these concepts, principles, and procedures to form a shared basis for food and feed measures taken in the Member State and at the Union level. Since that moment, the market perspective has been present in food law.

Besides laying down the principles on which modern food law should be based, the General Food Law also aims at establishing the European Food Safety Authority and procedures for food safety crises, including the so-called Rapid Alert System. The General Food Law cannot be considered as a code encompassing all the food legislations.[26] Nevertheless, it establishes a comprehensive framework for food and feed safety, covering different legislations such as animal welfare, hygiene, contaminants and residues, novel food, additives and flavourings. In short, the General Food Law's adoption laid the foundations for alignment with the principles encompassed in the Regulation itself, and for the "rationalisation" of the entire normative framework.[27]

For this book's aim, it is useful to outline the General Food Law's structure, focusing mostly on its main provisions. The scope of the description is to reach a better understanding of the regulatory architecture concerning the primary European Food Law's Regulation before the amendments introduced by the Regulation (EU)

---

[25] See Schebesta and Purnhagen (2024), p. 113 ff.

[26] Van der Meulen and Szajkovska (2020), p. 167.

[27] Santini (2013), p. 252. The author specifies that the Regulation "has been characterized by a tendency to concentrate in a small number of text the basic rules governing horizontal issues (that is, those issues that are not confined to specific food products), and on the other by the use of regulation instead of the pre-existing directives, in other words by an approach designed to bring about effective uniformity in this field, and not merely harmonisation of the rules of the Member States"(p. 252).

2019/1381 on the transparency and sustainability of the EU risk assessment in the food chain.[28] Later, the changes that have been included will be described.

4.2.1 Scope and Definitions

After the initial Recitals, the General Food Law in Chapter I points out the scope and the definitions.

Article 1 GFL emphasised that it serves as a foundation for ensuring a high level of protection of human health and consumers' interests. This includes addressing the diversity in the food supply chain and ensuring the functioning of the internal market. Article 1 also sets out common *principles* and *responsibilities*, and "the means to provide a strong science base, efficient organisational arrangements and procedures to underpin decision-making in matters of food and feed safety". The GFL takes a food chain approach[29] that encompasses all stages of "production, processing and distribution of food, and also of feed produced for, or fed to, food-producing animals" (Article 4). The network of steps and the variety of stakeholders involved in food production that result in this approach underline the idea that each actor, such as primary producers, processors or distributors, holds a portion of responsibility for ensuring food safety.[30]

Among all the definitions, Article 2 is of particular relevance. It stipulates that food—or foodstuff—means "any substance or product, whether processed, partially processed or unprocessed, intended to be, or reasonably expected to be ingested by humans".[31] It is also worth remembering the definition of food business operator (FBO) as "the natural or legal persons responsible for ensuring that the requirements of food law are met within the food business under their control".[32]

[28] Regulation (EU) 2019/1381 of the European Parliament and of the Council of 20 June 2019 on the transparency and sustainability of the EU risk assessment in the food chain and amending Regulations (EC) No 178/2002, (EC) No 1829/2003, (EC) No 1831/2003, (EC) No 2065/2003, (EC) No 1935/2004, (EC) No 1331/2008, (EC) No 1107/2009, (EU) 2015/2283 and Directive 2001/18/EC.

[29] Purnhagen (2019), p. 19.

[30] Vapnek (2007), pp. 989–990.

[31] "Food" includes "drink, chewing gum and any substance, including water, intentionally incorporated into the food during its manufacture, preparation or treatment". Several products are not considered "food", for instance feed and live animals unless they are prepared for placing on the market for human consumption as well as products such as medicines which are covered by a special regulatory regime. The Directive currently applicable to medicinal product is Directive 2001/83/EC which defines a medical product as "any substance or combination of substances presented for treating or preventing disease in human beings" or "any substance or combination of substances which may be administered to human modifying physiological functions in human beings", see Van der Meulen and Szajkovska (2020), p. 171.

[32] Article 3(3) General Food Law. According to Article 3(2), "food business" means "any undertaking, whether for profit or not and whether public or private, carrying out any of the activities related to any stage of production, processing and distribution of food".

4.2.2 *General Principles*

Articles 5 to 10 of the General Food Law are designed as general principles for all the Regulation, forming a "general framework of horizontal nature to be followed when measures are taken", as stated in Article 4 GFL.[33] Besides the general objectives, these principles encompass the risk analysis, the precautionary principle, consumers interests' protection, and the *principle of transparency*. All measures must take into account the general objectives outlined in Article 5 GFL, which are the high level of protection of *human life and health* and the protection of *consumers' interests*, the protection of animal health and welfare, plant health and the environment, and the free movement in the European Union of food and feed. Where applicable, international standards should also be considered when developing or adapting food law.[34]

4.2.2.1 The Risk Analysis Principle

To fulfil the general objectives outlined in Article 5, food law must be founded on risk analysis unless the circumstances or the nature of the measure make it inappropriate.[35] In Recital 17, the General Food Law prescribes that measures regulating food and feed issues should be based on risk analysis: a "systematic method" to analyse risks, decide and communicate upon how these risks will be handled.[36]

The European Union has played a leading role in the development of risk analysis principles and their international acceptance. In line with the framework of the GFL, food law is built upon three interrelated components of risk analysis: risk assessment, risk management and risk communication.[37] The White Paper on Food Safety, together with the Communication of the Commission on the precautionary principle introduced this tripartite structure, and the GFL strictly adheres to these documents in defining the risk analysis components. Actually, the need for a separation between the two components of risk assessment and risk management has been

---

[33] As the Court of Justice of the European Union (CJEU) stated in the case *Queisser Pharma* (Judgment of the Court (Fourth Chamber) of 19 January 2017, *Queisser Pharma GmbH & Co. KG v Bundesrepublik Deutschland*, C-282/15, ECLI:EU:C:2017:26) the general principles of food law form a general framework of horizontal nature to be followed when measures are taken even if these measures apply only to purely domestic cases. Hence, national measures must meet the requirements concerning the risk analysis and the precautionary principle. The food business operators, therefore, can seek judicial review of national measures also regarding the conformity with the general principles from Article 5 to 10 General Food Law. See Schebesta (2020); Hiller (2018), pp. 34–38.

[34] See Borghi (2003a), p. 191 ff., Schebesta and Purnhagen (2024), p. 118.

[35] Article 6 General Food Law.

[36] De Boer (2019b), p. 2.

[37] Article 3(10) General Food Law.

recognised at the international level by the Codex Alimentarius Commission. Its standards are considered a model for the European legal system.[38]

We will briefly define these components of the risk regulatory architecture within EU food law, as they are essential for the subsequent analysis. *Risk assessment* must be undertaken in an independent, objective and *transparent* manner based on the best available science.[39] The goal of risk assessment is to offer scientific advice which relies on extensive information gathering and analysis.[40] Risk assessment involves both determining when a risk is considered unacceptable and conducting a scientific evaluation of the risk.[41] In turn, risk assessment consists of four different steps: *hazard identification*, which identifies the nature of the adverse effects related to the properties of the substance; *hazard characterization*, which establishes the relationship among the level of a substance and the probability of an adverse health effect; the *exposure assessment*, which measures the amount of the substance that people are exposed to; and the *risk characterization*, which determines the probability of the adverse health effects on the basis of the previous step.[42] This process primarily focuses on evaluating and assessing biological, chemical, or physical aspects, although social considerations are also taken into account.[43] Risk assessment falls under the responsibility of the European Food Safety Authority which provides the necessary information to the public authorities, particularly the European Commission, to determine whether the risks are acceptable or whether to take appropriate measures to reduce them.[44] The distinction between risk assessment and risk management reflects the different roles and competencies of the European Food Safety Authority and the European Commission, ensuring that the scientific integrity of the risk assessment process is maintained by separating the phases to prevent political influence.[45] The distinction between risk assessment and risk management should then ensure that risk assessment remains free from any "preconceived opinions related to management solutions".[46] As stated by the Court of Justice of the Union, since the risk assessment is a "scientific process", the public

---

[38] The three components of risk analysis "should be applied within an overarching framework" for management of food related risks to human health, see Working Principles for Risk Analysis, Section IV of the Procedural Manual, FAO and WHO (2023), p. 99 ff. In the Decision of the 22nd Session of the Codex Alimentarius Commission, 1997, the Codex Alimentarius Commission has stated that "there should be a functional separation of risk assessment and risk management, while recognising that some interactions are essential for a pragmatic approach", see Gabbi (2007), p. 128. See Schebesta and Purnhagen (2024), p. 118, Fornabaio and Poto (2016), p. 316.

[39] Recital 18 General Food Law.

[40] White Paper on Food Safety, COM (1999) 719 final, par. 31, p. 14.

[41] Meyer (2006), p. 147.

[42] Article 3 (11) General Food Law. See De Boer (2019b), p. 2; WHO and FAO (2006).

[43] Van der Meulen and Szajkovska (2020), p. 184.

[44] Van der Meulen and Szajkovska (2020), p. 185.

[45] Fornabaio and Poto (2016), p. 317. See Gabbi et al. (2020), pp. 196–214.

[46] Gabbi (2007), p. 128.

authority must "entrust a scientific assessment to experts who will provide the scientific advice".[47]

Risk management involves weighing policy alternatives in light of risk assessment results and, if necessary, implementing appropriate measures to prevent, mitigate or eliminate the risk.[48] Food safety decisions cannot be based only on science. Establishing policy in a democratic way obliges to consider a "broad array" of values and preferences.[49] Thus, in the risk management phase, decision-makers need to consider a broad range of factors in addition to scientific risk assessment. These include, for instance, selecting the most effective risk reduction measures based on the specific stage of the food supply chain where the issues occur, assessing the feasibility of controlling risks, socio-economic effects, considering environmental impact or a wide range of other factors that legitimate the matter under consideration.[50] More in detail, these "other legitimate factors" are recognised as concerns of social, economic or ethical character, such as consumer risk perception, animal welfare, and traditional production methods.[51]

The question of how much discretion is conferred on public authorities reflects the extent to which risk managers can deviate from science in risk regulation.[52] The inclusion of socio-economic considerations in the decision-making process is an opportunity to promote diversity in the internal market by considering traditional, cultural or social implications as opposed to "technocratic" food safety regulation based on "sound science".[53]

In the risk management phase, as will be seen shortly, one may invoke the precautionary principle, where the conditions laid down in Article 7 GFL are met after the available information has been assessed, and in cases where scientific uncertainty persists.

Lastly, *risk communication* refers to the interactive exchange of information and opinions throughout the risk analysis. This process occurs among risk assessors, risk managers, consumers, feed and food businesses, academics, and other interested parties.[54] While Article 6 General Food Law does not explicitly address this component, we will see how Regulation (EU) 2019/1381 develops a more comprehensive explanation of this aspect of risk assessment.

---

[47] Judgement of the Court of First Instance (Third Chamber) 11 September 2002, *Pfizer Animal Health SA v Council of the European Union*, T-13/99, ECLI:EU:T:2002:209, paragraph 157.

[48] See Article 3 General Food Law.

[49] Szajkowska (2011), p. 529.

[50] Van der Meulen and Szajkovska (2020), p. 185.

[51] Szajkowska (2011), p. 523.

[52] Szajkowska (2011), p. 523.

[53] Szajkowska (2011), p. 538.

[54] Article 3(13) General Food Law.

4.2.2.2 The Precautionary Principle

Since science does not guarantee absolute certainty, the precautionary principle is applied when risk assessment fails to offer a scientific basis for the decision-making process and when uncertainties exist regarding the harmful effects of food on human health.[55] The precautionary principle "adds subtlety"[56] to the risk-analysis principle. In the specific circumstances where a risk to life or health exists but persists in a situation of uncertainty, the precautionary principle provides a mechanism to determine risk management measures or other actions in order to ensure the high level of health protection chosen by the European Union.[57] The principle appeared for the first time in 1992 in the EC Treaty's environmental title, but its scope was not limited to environmental issues.[58] Since then, it has become a legal obligation that science shall play a role in policy making through, among other things, the precautionary principle.[59] It might also be recalled that in several case law of the Court of Justice of the European Union, the judges applied the precautionary principle with regard to the validity of the European measures imposing the embargo on British beef exports in the wake of the BSE crisis.[60] The Court of Justice delivered this statement, repeated in subsequent rulings: "Where there is uncertainty as to the existence or extent of risk to human health, the institutions may take protective measures without having to wait until the reality and the seriousness of those risks become fully apparent".[61] The precautionary principle has faced significant criticism for being too vague or restrictive and is mentioned in Article 191 TFEU without further clarification.[62]

The key document to understand the precautionary principle is the European Commission's Communication on the Precautionary Principle. It explains that the principle "is relevant only in the event of a potential risk, even if this risk cannot be fully demonstrated or quantified or its effects determined because of the insufficiency or inclusive nature of the scientific data", but it cannot be used in any

[55] Fornabaio and Poto (2016), p. 316.

[56] Van der Meulen and Szajkovska (2020), p. 187.

[57] Recital 21 General Food Law. In the Communication from the Commission on the precautionary principle, 2.2.2000, COM (2000) 1 final, par. 2 the EC states that "(…) the precautionary principle is not defined in the Treaty, which prescribes it only once – to protect environment. But in practice its scope is much wider, and specifically where preliminary objective scientific evaluation, indicates that there are reasonable grounds for concern that the potentially dangerous effects on the environment, human, animal or plant health may be inconsistent with the high level of protection chosen for the Community".

[58] Van der Meulen and Szajkovska (2020), p. 187; De Vido (2020), p. 1324.

[59] Purnhagen (2015), p. 903.

[60] Van der Meulen and Szajkovska (2020), p. 189.

[61] See Case C-157/96, *The Queen v. Ministry of Agric., Fisheries and Food* (ECLI:EU:C:1998:191); Case C-180/96, *U.K. Gr. Brit and N. Ir. V. Commission* (ECLI:EU:C:1998:192); Case C-209/96, *U.K. Gr. Brit and N. Ir. V. Commission* (ECLI:EU:C:1998:448).

[62] Schebesta and Purnhagen (2024), p. 120.

circumstances to justify the adoption of arbitrary decisions.[63] Besides, the Communication provides that the precautionary measures shall be: (1) proportional to the chosen level of protection; (2) non-discriminatory in their application; (3) consistent with similar measures already in place; (4) based on an examination of the potential cost-benefit analysis of action or lack of action; (5) subject to review, in case of new scientific data; (6) shall facilitate the production of the scientific evidence necessary for a more comprehensive risk assessment.[64]

In the General Food Law, the precautionary principle is outlined in Article 7, and it deals with decision-making in cases of scientific uncertainties, stating also the conditions under which the principle is applied. The General Food Law invokes this principle to ensure health protection and sets a consistent basis for using the principle within the European Union. As mentioned, the Commission's Communication emphasised the necessity of a proportionality test, stating that precautionary measures "must not be disproportionate to the desired level of protection and must not aim at zero risk, something which rarely exists".[65] Similarly, Article 7 (2) General Food Law imposes that the measures taken under recourse to the precautionary principle "shall be proportionate and no more restrictive of trade than is required to achieve the high level of health protection" chosen in the Union, "regard being had to technical and economic feasibility and other factors regarded as legitimate in the matter under consideration".[66]

4.2.2.3 Consumers' Interests

As outlined in Article 8 of the General Food Law, the primary goal of food law is to protect consumers' interests, providing a basis for consumers to make informed choices in relation to the food they consume. Moreover, food law seeks to prevent fraudulent or misleading practices, food adulteration, and any other actions which may mislead the consumer.[67] The focus on consumer protection in Article 8 of Regulation (EC) No 178/2002 is primarily on safeguarding economic interests, even if related to food safety. A key term in the General Food Law associated with consumers' protection is "confidence". This concept relates to the decision-making process (as outlined in Recital 9) and the scientific basis of food law (in Recital 18)

---

[63] Communication from the Commission on the precautionary principle, 2.2.2000, COM (2000) 1 final, par. 5.1.

[64] Communication from the Commission on the precautionary principle, 2.2.2000, COM (2000) 1 final, p.3. See also Tosun (2013), p. 1519.

[65] Communication from the Commission on the precautionary principle, 2.2.2000, COM (2000) 1 final, par. 6.3.1.

[66] Article 7 adds that "the measures shall be reviewed within a reasonable period of time, depending on the nature of the risk to life or health identified and the type of scientific information needed to clarify the scientific uncertainty and to conduct a more comprehensive risk assessment".

[67] Article 8 General Food Law.

and represents a value that should be upheld in commercial relations.[68] By combining both Recitals 9 and 18 with Article 8, it becomes clear that the concept of confidence is integral to this regulation even though it is not explicit in the Article's text. Consumers are not only merely beneficiaries of the regulatory framework, but they also hold rights and obligations as food beneficiaries, ensuring that the product purchased meets the promised qualities, including food safety.[69]

4.2.2.4 Principles of Transparency

At the end of the part dedicated to the general principles of food law, Chapter II Section 2 of the General Food Law is titled "Principles of transparency" and contains Articles 9 and 10 GFL. Article 9 regards public consultation and it aligns with the indications that emerged in the White Paper on European Governance establishing a general principle of *governance* in the food sector. It imposes an "open and transparent public consultation, directly or through representative bodies, during preparation, evaluation and revision of food law".[70] Exceptions to this rule apply in urgent situations where consultations are not feasible. Aligning with the indications emerged in the White Paper on European Governance, Article 9 established a general principle of *governance* in the food sector.[71] The obligation to consult the public serves the two main objectives of maximising the use of available resources and information and fostering greater acceptance of the adopted rules by providing an opportunity to engage in the legislative process.[72] Article 10 mandates the provision of public information where there are "reasonable grounds to suspect that a food or feed may present a risk for human and animal health".[73] The information should be proportionate to the risk's nature, seriousness and extent, and public authorities shall take "appropriate steps" to inform the general public of the nature of the risk in relation to health. In doing so, they shall identify "to the fullest extent possible" the food or feed, or type of food or feed, the risk that may present and the measure to be taken to prevent, reduce or eliminate the risk.[74] Lastly, public information should be taken following European Union law and national law's provisions on

---

[68] Borghi (2003b), p. 222.

[69] Borghi (2003b), p. 222. The author points out that even if the consumers' legitimate expectations concern mainly economic interest, only a transparent legislation, reliable risk assessment and efficient risk communication would protect them. The system's credibility and its actual suitability to food safety requirements are basic needs. Without them, it would not make sense arising a problem of prevention against misleading business practices.

[70] Article 9 General Food Law.

[71] Viti (2003), p. 224.

[72] Conte-Salinas and Wallau (2016), p. 591.

[73] Article 10 General Food Law.

[74] Article 10 General Food Law.

access to documents. The Court of Justice of the European Union's case *Berger*[75] confirmed that while EU law does not mandate it, national law may allow public authorities to inform the public when a food is considered unsafe, also because it is unfit for human consumption.[76]

4.2.3 General Obligations of Food Trade

Articles 11 to 13 are designed as general obligations of food trade. Recital 23 asserts that the safety and confidence of consumers within the European Union are of the greatest importance.[77] The obligations on food trade concern the importers' and exporters' obligations to comply with the European Union food law unless the target country applies different laws.[78] Article 11 addresses the food and feed imported in the European Union, while Article 12 focuses on export and potential bilateral agreements concluded between the Union or one of its Member States and a third country.[79] Article 13 further requires that the EU will contribute to international standards development, where relevant and appropriate. Therefore, the EU promotes the alignment between international technical standards and food law, such as in the case of the Codex Alimentarius' concept of "food safety culture" now integrated into different EU food law instruments.[80]

4.2.4 General Requirements of Food Law

Section 4 of the General Food Law outlines the general regulatory requirements for food law (Articles 14 to 21). Article 14 of the General Food Law aims at protecting public health and the consumer from food that either poses a health risk or is deemed

[75] Judgment of the Court (Fourth Chamber), 11 April 2013, *Karl Berger v Freistaat Bayern*, C-636/11, ECLI:EU:C:2013:227.

[76] See Schebesta and Purnhagen (2024), p. 124.

[77] The European Union is considered as a major global trader in food and feed and, in this context, "it has entered into international trade agreements, it contributes to the development of international standards which underpin food law, and it supports the principle of free trade in safe feed and safe, wholesome food in a non-discriminatory manner, following fair and ethical trading practices" (Recital 23).

[78] Purnhagen (2019), p. 20.

[79] See Guidance on the implementation of Articles 11, 12, 14, 17, 18, 19 and 20 of Regulation (EC) No 178/2002 on General Food Law – Conclusion of the Standing Committee on the Food Chain and Animal Health, revisited version of 26 January 2010 (available at https://food.ec.europa.eu/system/files/2016-10/gfl_req_guidance_rev_8_en.pdf/) para. VII.

[80] Schebesta and Purnhagen (2024), p. 125.

unacceptable,[81] and responds to the need of establish a general requirement to ensure that only safe food is placed on the market.[82] More precisely, Article 14 expresses the general principle according to which "food shall not be placed on the market if it is unsafe". Namely, this condition occurs whether food is either *injurious* to health or *unfit* for human consumption, depending on whether there are requirements concerning the effects or the quality of the food itself. Furthermore, the law provides the definitions and the factors to take into account in determining the unsafety of the food.[83]

Articles 17 and 19 place on the food business operator (FBO) a primary legal responsibility of ensuring compliance with food law,[84] while Member States are responsible for enforcing, monitoring, and verifying these obligations' fulfilment.[85] In the responsibilities system, we can note that businesses must make sure that food on the market is not *unsafe*, rather than provide safe food or justify food's safety.[86] Once it has been ascertained that the food is unsafe, the businesses have an obligation to *withdraw* the food in question. If the product has already reached the consumer, to *inform* "effectively and accurately" the consumers and recall the products already supplied.[87]

Article 16 General Food Law lays down another rule for consumer protection: the presentation of food, including advertising and labelling, shall not mislead consumers. In addition, food business operators must adhere to the traceability

---

[81] See Guidance on the implementation of Articles 11, 12, 14, 17, 18, 19 and 20 of Regulation (EC) No 178/2002 on General Food Law, para .3.3.

[82] Recital 27 General Food Law.

[83] Article 14 provides a catalogue of these "unsafety" definitions. Due to the importance of this norm, we must recall them. Article 14 (3) states that "in determining whether any food is unsafe, regard shall be had: (a) to the normal conditions of use of the food by the consumer and at each stage of production, processing and distribution, and (b) to the information provided to the consumer, including information on the label, or other information generally available to the consumer concerning the avoidance of specific adverse health effects from a particular food or category of foods". According to Article 14(4), instead, the food is *injurious* to health having regard "(a) not only to the probable immediate and/or short-term and /or long-term effects of that food on the health of a person consuming it, but also on subsequent generations; (b) to the probable cumulative toxic effects; (c) to the particular health sensitivities of a specific category of consumers where the food is intended for that category of consumers". Article 14(5) requires that for classifying a food as *unfit* for human consumption, "regard shall be had to whether the food is unacceptable for human consumption according to its intended use, for reasons of contamination, whether by extraneous matter or otherwise, or through putrefaction, deterioration or decay". For a detailed analysis of Article 14, see Van der Meulen (2012), pp. 117–125.

[84] See Guidance on the implementation of Articles 11, 12, 14, 17, 18, 19 and 20 of Regulation (EC) No 178/2002 on General Food Law, para II.

[85] Article 17(2) General Food Law. For this aim, Article 17(3) states that "they shall maintain a system of official controls and other activities as appropriate to the circumstances, including public communication on the food and feed safety risk, food and feed safety surveillance and other monitoring activities covering all stages of production, processing and distribution".

[86] See Purnhagen (2019), p. 21 and Purnhagen (2015), p. 916; Santini (2013), pp. 257–258.

[87] Article 19 General Food Law.

requirements set out in Article 18 General Food Law. This rule includes the possibility to identify each supplier in the food chain and to communicate through labelling and relevant documentation or information.[88] The establishment of a traceability system prevents the risk that the internal market's functioning could be jeopardised by the impossibility of tracing food.[89]

Lastly, even in the General Food Law framework, the EU rules concerning product liability remain applicable.[90]

4.2.5 The European Food Safety Authority

Chapter III of the General Food Law is devoted to the Food Safety Authority (EFSA or "the Authority").[91] The General Food Law implements the provisions of the White Paper on Food Safety by transferring the duties of the previous scientific Committees to a regulatory agency. In the White Paper, the aim of restoring consumer confidence is connected with establishing an Authority "with particular responsibilities for both risk assessment and communication on food safety issues".[92]

Article 22 sets out the Authority's *mission*. EFSA's main task is to provide scientific advice and scientific and technical support for the EU legislation and policies in all fields that directly or indirectly affect food and feed safety. The General Food Law clearly defines EFSA's role as limited to food safety, highlighting the distinction from risk management. Recital 35 GFL in this regard specifies that the Authority "should be an *independent* scientific source of advice, information and risk

---

[88] Article 18(4) General Food Law. EFSA is an EU Agency, therefore it follows the doctrinal and judicial debate about the EU Agencies' delegation of powers. On the famous *Meroni* (Judgment of the Court of 13 June 1958, *Meroni & Co., Industrie Metallurgiche, SpA v High Authority of the European Coal and Steel Community*, C-9/56, ECLI:EU:C:1958:7) and *Romano* (Judgment of the Court (First Chamber) of 14 May 1981, *Giuseppe Romano v Institut national d'assurance maladie-invalidité*, C-98/80, ECLI:EU:C:1981:104) delegation doctrine and the subsequent "new delegation doctrine" in *Short selling* case (Judgment of the Court (Grand Chamber), 22 January 2014, *United Kingdom of Great Britain and Northern Ireland v European Parliament and Council of the European Union*, C-270/12, ECLI:EU:C:2014:18), see Scholten and van Rijsbergen (2014), pp. 389–405; Chamon (2011), p. 1055 ff.; Alberti (2018), p. 274 ff.; Simoncini (2020). Particularly, in *Short selling* the CJEU emphasised the technical expertise of the European Securities and Markets Authority (ESMA). The topic of transparency in that case law concerns the transparency in financial markets both to the regulators and to the public. See Maijoor (2011), p. 9; Mendes (2017), pp. 443–472; Chamon (2023). See also Chap. 5, para. 5.1.

[89] Recital 28 General Food Law states that it is important to establish a "comprehensive system of traceability within food and feed businesses so that targeted and accurate withdrawals can be undertaken or information given to consumers or controls official, thereby avoiding the potential for unnecessary wider disruption in the event of food safety problems".

[90] Article 21 General Food Law. See also Santini (2013), p. 257.

[91] On EFSA, see Alemanno and Gabbi (2016).

[92] White Paper on Food Safety, COM (1999) 719 final, Brussels 12 January 2000, Chap. 5, para. 29.

communication in order to improve consumer confidence".[93] Therefore, it should ensure constant, competent and independent technical advice and scientific support in the matter of food safety by collecting and analysing data. The Authority's mission, then, should include providing scientific advice and technical support and scientific opinions.[94] EFSA acts as an essential reference for the new system supporting the preparation, definition and implementation of food law.[95] The Authority must provide scientific and technical assistance, advice, and opinions on relevant matters, and gather information and analysis by collecting and summarising scientific and technical data within its scope (Article 23). To enhance its work, EFSA must collaborate with different organisations to build networks with national agencies and scientific bodies. Moreover, the Authority must identify and communicate emerging risks.[96]

EFSA is composed of four main bodies with specific tasks and duties. Particularly, the Management Board has the duty to ensure the Authority's operation by preparing the work schedule and reporting the activities.[97] It shall ensure that the Authority carries out its mission and performs the assigned tasks under the conditions laid down in the General Food Law.[98] The Executive Director is the Authority's legal representative and has various duties, among which we should recall the proposal for the Authority's work programmes in consultation with the Commission; the implementation of the programmes and decisions adopted by the Management Board; ensuring the provisions of appropriate scientific, technical and administrative support for the Scientific Committee and the Scientific Panels and that the Authority carries out its tasks in accordance with the requirements of its users.[99]

The Advisory forum comprises representatives from the competent bodies in the Member States that undertake EFSA's similar tasks to constitute a mechanism of exchanging information and ensuring cooperation about the networking system.[100]

---

[93] Emphasis added.

[94] It is worth recalling Article 22 (5), which includes in the mission of the Authority: "(a) scientific advice and scientific and technical support on human nutrition in relation to Community legislation and, at the request of the Commission, assistance concerning communication on nutritional issues within the framework of the Community health programme; (b) scientific opinions on the other matters relating to animal health and welfare and plant health; (c) scientific opinions on products other than food and feed relating to genetically modified organisms as defined by Directive 2001/18/EC and without prejudice to the procedure established therein".

[95] See Rusconi (2016), pp. 467 ff.

[96] Recital 50 specifies that is necessary to assign to the Authority "anticipatory task of collecting information and exercising vigilance and providing evaluation of and information on emerging risks with a view to their prevention". The aim is disposing preventive instruments in the long term and improving the identification of emerging risks.

[97] Article 25 General Food Law. See Rusconi (2016), p. 467 and Gabbi et al. (2020), p. 200.

[98] Article 25 (7) General Food Law.

[99] Article 26 General Food Law lists the other responsibilities of the Executive Director, e.g. the preparation of the statement of revenue and expenditure and the execution of the budget of the Authority and the development of contact with the European Parliament.

[100] Recital 44 and Article 27 General Food Law.

Scientific Committee and Permanent Scientific Panels are responsible for providing scientific opinions, respectively within their own competencies.[101] The Scientific Committee consists of the Scientific Panels' chairs complemented by other six independent experts who do not belong to any of the Scientific Panels.[102] Scientific opinions are the main tools through which the Authority's activity is carried out. Through them, the *risk assessment*—a core aim of the system—is achieved.[103] The Authority shall give its scientific opinion at the request of the Commission, of the European Parliament or of a Member State (and in all cases where EU law provides for the consultation of the Authority) or, on matters falling within its mission, on its initiative.[104] It can also be requested by the Commission to provide scientific or technical assistance.[105] In pursuing its duties, the Authority may commission scientific studies "in an open and *transparent* fashion".[106]

Since EFSA's main task focuses on risk assessment, the General Food Law lays down specific rules to identify emerging risks and the receipt of messages forwarded via the rapid alert system.[107]

Section 4 of Chapter III enshrines the requisites of independence (art. 37) and *transparency* (art. 38). These are the "operating principles"[108] which are also highlighted by the White Paper on Food Safety.[109]

Articles 39 to 42 lay down other rules regarding confidentiality, communications from the Authority, access to documents and contacts with consumers, producers and other interested parties.

---

[101] Article 28 General Food Law.

[102] Article 28 (3) General Food Law. The Scientific Panels corresponding to different areas, are: the Panel on food additives and flavorings (FAF); the Panel on additives and products or substances used in the animal feed (FEEDAP); the Panel on plant health (PLH); the Panel on Plant Protection Products and their Residues (PPR); the Panel on genetically modified organisms (GMO); the Panel on dietetic products, nutrition and allergies (NDA); the Panel on microbiological hazards (BIOHAZ); the Panel on contaminants in the food chain (CONTAM); the Panel on animal health and welfare (AHAW); the Panel on Food Contact Materials (FCM).

[103] Rusconi (2016), p. 468.

[104] Article 29 General Food Law. Pursuant to Article 29 (2), requests shall be accompanied by background information explaining the scientific issue to be addressed and the Community interest. In case of divergences between EFSA's scientific opinions and the scientific opinions issued by other bodies, Article 30 General Food Law provides a mechanism to remedy. Namely, EFSA in case of divergence shall contact the body to ensure that "all relevant scientific information is shared" and in order "to identify potentially contentious scientific issues". EFSA and the other body must cooperate with the scope to resolve the divergence or preparing a joint document to the Commission clarifying the contentious scientific issues and identifying the relevant uncertainties in the data.

[105] Article 31 General Food Law.

[106] Article 32 General Food Law.

[107] Article 34 and Article 35 General Food Law. On Article 34 General Food Law, see also Kocharov (2010), p. 144.

[108] Recital 56 General Food Law.

[109] White Paper on Food Safety, COM (1999) 719 final, Brussels 12 January 2000, Chapter 4, para. 41–43.

Finally, in Articles 43–49 General Food Law pose other final rules concerning financial provisions and general provisions on liability and participation of third countries.

4.2.6 *The Rapid Alert System, Crisis Management and Emergencies*

Articles 50–60 General Food Law set up the Rapid Alert System for food and feed (RASFF) and other tools designed to manage a crisis. The RASFF is an instrument to communicate food and feed risks. It has been created to allow authorities to be rapidly informed of the risks and take any necessary action. In particular, the RASFF is a tool for communication between authorities in different Member States, the European Commission and some other countries which have joined the system, such as Norway, Liechtenstein and Iceland.[110] Even if it is designed as an internal means of communication, it has serious external effects due to the nature of the information transmitted through it—concerning food-related health problems—and the fact that it has to reach many recipients in different countries.[111]

Moreover, Articles 53 and 54 General Food Law set out other different rules regarding emergency measures for food and feed originating in the Union or imported from a third country.[112] Besides, the Commission shall draw up, in close cooperation with the Authority and the Member States, a general plan for crisis management in the field of safety of food and feed.[113] The general plan shall specify the situation that cannot be solved by applying Articles 53 and 54. Finally, Articles 56 and 57 concern the establishment of a crisis unit set up by the Commission and in which the Authority shall participate and provide scientific and technical assistance.

Lastly, Chapter VI General Food Law contains procedures and final provisions.

[110] See Fuentes (2017), p. 121; Santini (2014), pp. 1433–1444; D'Addezio and Maccioni (2016), p. 301 ff. and Banati and Klaus (2010), pp. 10–21.

[111] Fuentes (2017), p. 121.

[112] Article 53 General Food Law lists a series of measures, among which we recall the suspension of the placing on the market or use of the food or feed in question, the provision of special conditions for the food or feed in questions or any other appropriate interim measure.

[113] Article 55 General Food Law.

4.3 The Fitness Check of the General Food Law

In June 2019, the General Food Law reform was approved, that is Regulation (EU) 2019/1381 on the transparency and sustainability of the EU risk assessment in the food chain.[114] We refer to this Regulation also with the expression "New Transparency Rules", "Transparency Regulation", or "the reform". This Regulation has been applicable since 27 March 2021. The Regulation (EU) 2019/1381 considers the result of the "Fitness Check" of the General Food Law, launched in 2014.

More in general, Fitness Checks derive from the review of the Union Legislation under the Regulatory Fitness and Performance Programme (REFIT). The REFIT is a process affecting the whole policy cycle—from the design of legislation to implementation, enforcement, evaluation and revision—aiming at ensuring that European Union laws reach their objectives at minimum costs for the citizens' and businesses' benefits.[115] Fitness Checks, thus, are policy evaluations aiming at assessing if a regulatory framework reaches its purpose. They provide an evidence-based analysis of the extent to which EU actions are proportionate to their goals.[116] The Commission in a 2013 Working Document in which the comprehensive food sector's legislation was reviewed and assessed chose to evaluate the General Food Law under the REFIT.[117] In January 2018 the Fitness Check on the General Food Law was finalised.[118] According to the final report, the General Food Law maintains both aims of protecting human health and consumers' interests and the effective functioning of the internal market. It also resulted that the GFL contributes to the development of a level playing field throughout the European market. Among other factors, food safety, as recognised at the European level, increased the European food and drink industry's competitiveness.[119]

---

[114] Regulation (EU) 2019/1381 of the European Parliament and of the Council of 20 June 2019 on the transparency and sustainability of the EU risk assessment in the food chain and amending Regulations (EC) No 178/2002, (EC) No 1829/2003, (EC) No 1831/2003, (EC) No 2065/2003, (EC) No 1935/2004, (EC) No 1331/2008, (EC) No 1107/2009, (EU) 2015/2283 and Directive 2001/18/EC.

[115] See Communication from the Commission to the European Parliament, the Council, the European Economic and Social Committee and the Committee of the Regions—EU Regulatory Fitness, Brussels 12.12.2012, COM (2012) 746 final.

[116] See COM (2012) 746 final and https://ec.europa.eu/food/safety/general_food_law/fitness_check_en

[117] Commission Staff Working Document. A fitness check of the food chain—State of play and next steps, Brussels 5.12.2013, SWD (2013) 516 final. See also De Boer (2019b), p. 3.

[118] Commission Staff Working Document, The Refit evaluation of the General Food Law (Regulation (EC) No 178/2002), Brussels 15.1.2018, SWD (2018) 38 final. See also Ferrari (2019), p. 67.

[119] De Boer (2019b), p. 3. The Commission Staff Working Document, The Refit evaluation of the General Food Law, points out that "the GFL Regelation as a framework legislation providing one common and consistent basis both at EU and national level for the entire food chain combined with a great degree of harmonisation in other EU secondary food legislation has contributed to the effective functioning of the internal market by creating a level playing field for all feed and food

Nevertheless, the General Food Law evaluation identified several shortcomings, including the transparency of the risk analysis process. These shortcomings remain a "sensitive issue in terms of perception",[120] mainly in the risk assessment and the risk communication areas.

Moreover, the European citizens' initiative entitled "Ban glyphosate and protect people and the environment from toxic pesticides"[121] further supported concerns regarding transparency of the studies commissioned by the industry and submitted in the authorisation procedure. Glyphosate is an *active substance*[122] used in several pesticides to combat weeds affecting crops. Regulation (EC) No 1107/2009 (also called Regulation on plant protection products—PPPs)[123] foresees the authorisation procedure for the placing on the market of plant protection products. This Regulation sets out a prior authorisation system according to which substances can be placed on the market only if approved or authorised. More in detail, the approval criteria are laid down in Article 4 Regulation (EC) No1107/2009. The burden of proof is on the applicant who must establish that the active substances or PPPs do not have harmful effects on human health or unacceptable effects on the environment.[124]

Glyphosate became highly controversial because of conflicting risk assessments on its carcinogenicity in the context of the renewal of its approval in the EU. While the UN International Agency for Research on Cancer (IARC) considered that glyphosate should be classified "probably a carcinogen", EFSA concluded that "glyphosate is unlikely to pose a carcinogenic hazard to humans".[125] In the public debate EFSA has been accused of basing its evaluation on biased evidence, as highlighted by the ECI and two EU Parliament Resolutions.[126] The procedure took place between 2012 and 2017 and culminated in a first renewal of the approval for 5 years.

business operators (FBOs) in the EU market while reducing disruptions of trade where problems have occurred" (at p. 114).

[120] The Commission Staff Working Document, The Refit evaluation of the General Food Law, p. 115.

[121] See the European Citizen's initiative "Ban glyphosate and protect people and the environment from toxic pesticide", available at: https://europa.eu/citizens-initiative/initiatives/details/2017/000002_en. See also the Communication from the Commission on the European Citizens' Initiative "Ban glyphosate and protect people and the environment from toxic pesticides", Strasbourg, 12.12.2007, C (2017) 8414 final. See also Pichler and Kaufmann (2010); Karatzia (2017), pp. 177–208; Weisskircher (2020), pp. 797–815 and Santini (2019), pp. 187–189.

[122] Active substances are active components operating against pests and plant diseases.

[123] Regulation (EC) No 1107/2009 of the European Parliament and of the Council of 21 October 2009 concerning the placing of plant protection products on the market and repealing Council Directives 79/117/EEC and 91/414/EEC, OJ L 309, 24.11.2009, p. 1–50.

[124] Regulation 1107/2009 lays down a centralized EU-wide approval procedure, while a separate authorisation procedure takes place at the Member States level. On the EU Pesticide Authorisation procedure see Paulini (2020), pp. 481–497.

[125] On the dispute in the assessments made by IARC and EFSA, see Bozzini (2020) pp. 197–213, and Gabbi et al. (2020), pp. 211–212. For the EFSA study, see EFSA (2015). More in detail on the "glyphosate saga", see Arcuri and Hale Hendlin (2019); Donati (2020); Morvillo (2020); Röttger-Wirtz (2020); Capelli and Giangiobbe (2021).

[126] European Parliament resolution of 13 September 2018 on the implementation of the Plant Protection Products Regulation (EC) No 1107/2009 (2017/2128(INI)), P8_TA (2018)0356 and

The most recent assessment that led to the renewal of approval of glyphosate was carried out between 2019 and 2023 by Member States' Competent Authorities, the European Food Safety Authority (EFSA) and the European Chemicals Agency (ECHA) and did not identify scientific or legal justification for a ban.[127]

The debate around aspartame has also contributed to increase a climate of suspicion on EFSA's risk assessment even before the glyphosate *affair*. In this regard, the Authority's decision-making process has been criticised for potential commercial conflict of interests.[128] As noted in Recital 27 of the Regulation (EU) 2019/1381, the reform also addressed the glyphosate dispute. Following these Fitness Check's results and as a response to the European Commission's commitments regarding the ECI's initiative, the European Commission set up an Open public consultation.[129] A total of 471 participants, among which 318 citizens and 153 stakeholders, replied to the European Commission and submitted their opinions regarding the transparency of the risk assessment conducted by the European Food Safety Authority and based on industry studies, the risk communication, and the involvement of the Member States in EFSA's governance.[130] The European legislator then drafted a proposal on the transparency and sustainability of the EU risk assessment in the food chain following the evaluation and public consultations' outcomes.[131]

European Parliament resolution of 16 January 2019 on the Union's authorisation procedure for pesticides (2018/2153(INI)), P8_TA (2019)0023.

[127] See EFSA et al. (2023). See also Commission Implementing Regulation (EU) 2023/2660 of 28 November 2023 renewing the approval of the active substance glyphosate in accordance with Regulation (EC) No 1107/2009 of the European Parliament and of the Council and amending Commission Implementing Regulation (EU) No 540/2011, C/2023/8101, OJ L, 2023/2660, 29.11.2023.

[128] On the aspartame controversy, see Huizing Edinger (2014).

[129] The Open public consultation was open from the 23rd of January to the 20th of March 2018.

[130] See Commission Staff Working Document, Synopsis Report Accompanying the document Regulation on the European Parliament and the Council on the transparency and sustainability of the EU risk assessment in the food chain amending Regulation (EC) No 178/2002 [on General Food Law], Directive 2001/18/EC [on the deliberate release into the environment of GMOs], Regulation (EC) No 1829/2003 [on GM food and feed], Regulation (EC) No 1831/2003 [on feed additives], Regulation (EC) No 2065/2003 [on smoke flavourings], Regulation (EC) No 1935/2004 [on food contact materials], Regulation (EC) No 1331/2008 [on the common authorisation procedure for food additives, food enzymes and food flavourings], Regulation (EC) No 1107/2009 [on plant protection products] and Regulation (EC) No 2015/2283 [on novel foods], Brussels 11.4.2018, SWD (2018) 97 final. See also De Boer (2019b) p. 3. See Tosun and Varone (2020), Chearnaigh (2021), p. 703.

[131] Proposal for a Regulation of the European Parliament and of the Council on the transparency and sustainability of the EU risk assessment in the food chain amending Regulation (EC) No 178/2002 [on General Food Law], Directive 2001/18/EC [on the deliberate release into the environment of GMOs], Regulation (EC) No 1829/2003 [on GM food and feed], Regulation (EC) No 1831/2003 [on feed additives], Regulation (EC) No 2065/2003 [on smoke flavourings], Regulation (EC) No 1935/2004 [on food contact materials], Regulation (EC) No 1331/2008 [on the common authorisation procedure for food additives, food enzymes and food flavourings], Regulation (EC) No 1107/2009 [on plant protection products] and Regulation (EC) No 2015/2283 [on novel foods], Brussels 11.4.2018, SWD (2018) 179 final.

In June 2019 the Regulation (EU) 2019/1381 was adopted following the opinions on the Commission proposal of April 2018 by the Economic and Social Committee and the European Committee of the Regions, the subsequent discussions within the Council and its preparatory bodies, and the first readings of the European Parliament.

The Regulation aims to a new design of the EU risk assessment's transparency and sustainability, introducing and amending several provisions of the General Food Law. Furthermore, it amends various sectoral acts of secondary legislation. In the following part, we will explain the changes that have been introduced by the "Transparency Regulation". This part will focus only on the introductions and amendments occurring on the General Food Law. Later, we will conduct a conceptual analysis of the various obligations arising from the new rules with the scope of linking them with the dimensions of transparency discovered in Chap. 2 of this book.[132]

4.4 The Regulation (EU) 2019/1381 on Transparency and Sustainability of the EU Risk Assessment

The Regulation (EU) 2019/1381 introduced several amendments into the General Food Law. The key changes concern the risk communication, the composition of EFSA, the adjustments on risk assessment, and measures to ensure transparency and confidentiality. Even in the title of the Regulation, the focus is on the transparency and sustainability of risk assessment in the food chain. However, the term 'sustainability' appears to be unrelated to the 'sustainable development' concept. 'Sustainability' in the new Regulation seems to refer only to the governance changes within EFSA instead of having an environmental connotation. In Recital 18, the reference on the term sustainability concerns the Authority's expertise to ensure its "efficient operation". Even preparatory works and factsheets omit references on sustainable development as one of the EU's objectives, as expressed in Article 3 of the Treaty of the European Union (TEU). Since sustainability is a "multidimensional concept",[133] the interpretation suggested for this analysis is to intend it as "institutional" sustainability.

[132] See Chap. 2, para 2.

[133] On "unresolved ambiguity of food sustainability", see Schebesta and Candel (2020), p. 586.

4.4.1 Risk Communication

The REFIT evaluation found that the risk communication is generally not considered effective (Recital 3), affecting consumers' confidence in the outcome of the risk analysis procedure. Therefore, Art. 1 New Transparency Rules introduced a paragraph in Article 6 General Food Law stating that risk communication must fulfil the objectives and respect the general principles outlined in the new Article 8a and Article 8b.[134] Article 8a lays down several objectives of risk communication. We will be focusing on the main. To ensure transparency and strengthen the citizens' trust,[135] risk communication should "raise awareness and understanding of the specific issues under consideration",[136] including the entire risk analysis process and the differences in scientific assessment. The risk communication should foster a participatory and open dialogue between all interested parties to ensure that the risk analysis process is based on accuracy, comprehensiveness, *transparency*, consistency, and accountability.[137] In doing so, risk communication should clarify the risk management decisions by presenting the findings and their use, in combination with the other legitimate factors.[138] In this exposition, risk communication should also explain the reasons for the factors considered by the risk managers and how those factors were weighed up against each other.[139]

The subjects involved in the open dialogue and indicated in Article 8a are consumers, feed and food businesses, the academic community and all other interested parties.[140] In respect of these parties, the risk communication aims to ensure a fair and *transparent* exchange of information.[141]

Risk communication should inform the consumer about risk prevention strategies and help combat the spread of false information.[142]

Article 8b outlines the general principles of risk communication which recalls the principle of transparency and the exchange of transparent information among

[134] In Recital 8 Regulation (EU) 2019/1381 the necessity to establish general objectives and principles of risk communication should take into account the respective roles of risk assessors and managers, guaranteeing also their independence.

[135] Recital 4 Regulation (EU) 2019/1381.

[136] Article 1(1) Regulation (EU) 2019/1381 inserting Article 8a lett. (a) General Food Law: "raise awareness and understanding of the specific issue under consideration, including in cases of divergences in scientific assessment, during the entire risk analysis process".

[137] Recital 4 Regulation (EU) 2019/1381.

[138] Recital 5 Regulation (EU) 2019/1381. Article 8a lett. (b), (c), and (d) indicate, respectively, the objectives of "ensure consistency, transparency and clarity in formulating risk management recommendations and decisions"; "provide a sound basis, including, where appropriate, a scientific basis, for understanding risk management decisions"; "improve the overall effectiveness and efficiency of the risk analysis".

[139] Recital 5 Regulation (EU) 2019/1381.

[140] Article 8a lett. (f) General Food Law.

[141] Article 8a lett. (g) General Food Law.

[142] Article 8a lett. (h) and (i) General Food Law.

the interested parties to foster dialogue, including those not directly involved in the risk analysis process or not having a scientific background.[143]

The transparency reform introduces a "general plan for risk communication" in Article 8c, designed to achieve the objectives set out in Article 8a and in conformity with the general principles of risk communication in Article 8b. The general plan should be established by the Commission in cooperation with EFSA and Member States. It should promote an "integrated risk communication framework" for all risk assessors and risk managers at Union and national levels.[144] The general plan should identify the key factors to consider when implementing the risk communication activities, such as the level, nature, urgency and impact of the risk,[145] as well as the tools and channels to be used for the involvement of the interested parties.[146]

4.4.2 The Amendments on the Governance of EFSA

Article 25 and Article 28 General Food Law introduce further changes on the institutional framework of the European Food Safety Authority. The objective to increase the transparency of the risk assessment process is linked to the aim to strengthen the confidence of the general public in the risk analysis and in EFSA's independence. Article 25 reshapes the Management Board's governance. Before, the Management Board was composed of members appointed by the Council in consultation with the European Parliament from a list drawn up by the Commission. Now, Article 25 adds representatives of all Member States, of the European Parliament and of the Commission as well as of civil society and industry organisations.[147] Besides being experts in the fields of food chain law and policy, the representatives should demonstrate to have broad experience and expertise in the managerial, administrative, financial and legal fields.[148] Moreover, the new rules establish appointment mechanisms with the aim of securing the highest standards of competence.[149]

---

[143] Article 8b General Food Law.

[144] Recital 9 Regulation (EU) 2019/1381.

[145] Article 8c (2) and Recital 10 Regulation (EU) 2019/1381.

[146] Article 8c General Food Law.

[147] Recital 13 Regulation (EU) 2019/1381. Paragraph 1 of Article 25 General Food Law provides that "Each Member State shall nominate a member and an alternate member and an alternate member as its representatives to the Management Board. The members and alternate member thus nominated shall be appointed by the Council and have the right to vote".

[148] Recital 14 Regulation (EU) 2019/1381.

[149] In Article 25 General Food Law, Paragraph 1a regulates the number and the appointment mechanism of the other members. In accordance with Paragraph 2, the members shall be in charge for 4 years and may be renewed. The Management Board, as laid down in Paragraph 5, shall act by a majority of its members. Alternate members shall represent the members in their absence and vote on their behalf".

Instead, Article 28 introduces some changes in the Scientific Committee and in the Scientific Panels. Among the shortcomings resulting from the Fitness Check of the General Food Law was the Authority's capability to maintain a high-level expertise, partly due to a decrease in the number of candidates applying for appointments to the Panels.[150] Consequently, the reform aims to modify the selection and the appointment mechanism of the experts based on Member State proposals,[151] assigning an active role at the Executive Director.[152] It is essential to ensure that scientists are selected from among experts who actively conduct research and meet stringent independence criteria.[153] Moreover, the new rules require that EFSA supports the Scientific Committee and Scientific Panels' tasks by organising the preparatory work, including by drafting preparatory scientific opinions to be peer-reviewed and adopted by those Scientific Panels.[154]

4.4.3 The Adjustment to the Risk Assessment

The Transparency Regulation expanded EFSA's mandate with several amendments. In the General Food Law system, food business operators willing to place certain products on the market, such as GMOs and additives, or marketing other foodstuffs in particular novel foods, have to seek the authorisation from the European Commission. EFSA carries out the risk assessment, whereas businesses must submit scientific data and documents supporting their request.

Recital 19 of the Transparency Regulation recalls the principle for the applicant or the notifier of an authorisation procedure to prove that the subject of the application or the notification complies with the law.[155] The prior approvals or marketing

---

[150] Recital 16 Regulation (EU) 2019/1381.

[151] Article 28, Paragraph 5 General Food Law.

[152] Article 28, Paragraph 5 General Food Law requires that "The members of the Scientific Committee who are not members of the Scientific Panels and the members of the Scientific Panels shall be appointed by the Management Board, acting upon a proposal from the Executive Director, for a five-year term of office, which may be renewed, following publication of a call for expression of interest in the Official Journal of the European Union, in relevant leading scientific publications and on the Authority's website. The Authority shall publish such a call for expression of interest after having informed the Member States about the necessary criteria and fields of expertise". The procedure is set up in Paragraph 5a and 5b.

[153] Among the criteria, Article 28 Paragraph 5a requires "(i) a high level of scientific expertise; (ii) independence and absence of conflict of interests in accordance with Article 37(2) and the Authority's independence policy and implementation of that policy in respect of the members of the Scientific Panels; (iii) meeting the needs for the specific multi-disciplinary expertise. The Management Board shall ensure that the broadest possible geographical distribution is achieved in the appointments".

[154] Article 28, Paragraph 5e General Food Law. For some insights on the new EFSA governance' rules, see Leone (2021), pp. 536–568, Vos et al. (2023), p. 31.

[155] The authorisation procedures in food law are required for several products, such as GMOs, novel food and additives, enzymes, flavourings and extraction solvents. This principle of prior

authorisation requirements on products reverse the burden of proof by treating the products as dangerous unless the businesses demonstrate their safety. It has been argued that when the burden of proof is on the applicant or the notifier, human or animal health and the environment are better protected, rather than imposing on the Authority to prove that the subject matter is unsafe.[156] Therefore, the new rules are based on these premises. To support the application or the notification, applicants and notifiers should provide and submit relevant studies, including tests, to demonstrate the safety and efficacy of the subject matter.[157]

In this vein, it is essential that the request for authorisation or the notification submitted to the Authority complies with the requirements. Since the understanding of those specific requirements can be unclear for the applicants, especially if they are small and medium enterprises (SMEs), the Transparency Regulation provides a new tool. Namely, upon request, the Authority should offer a "pre-submission" advice related to the relevant legal requirements and content expectations.[158] Article 32a specifies that such advice must be "without prejudice and non-committal as to any subsequent assessment of application and notification". To ensure impartiality, the amended provision prevents the EFSA personnel from undertaking any preparatory duties related to the application.[159]

Article 32b sets out the requirement to notify studies commissioned or conducted to support applications or notifications. Both business operators and laboratories are required to notify the title and the scope of any studies.[160] In this way, information

---

approval is explained also in the Communication from the Commission on the precautionary principle, 2.2.2000, COM (2000) 1 final, para. 6.4.: "the principle of prior approval (positive list) before the placing on the market of certain products, such as drugs, pesticides or food additives. This is one way of applying the precautionary principle, by shifting the responsibility for producing scientific evidence. This applies in particular to substances deemed "a priori" hazardous or which are potentially hazardous at a certain level of absorption. In this case the legislator, by way of precaution, has clearly reversed the burden of proof by requiring that the substances be deemed hazardous until proven otherwise. Hence it is up to the business community to carry out the scientific work needed to evaluate the risk. As long as the human health risk cannot be evaluated with sufficient certainty, the legislator is not legally entitled to authorize use of the substance, unless exceptionally for test purposes".

[156] Recital 19 Regulation (EU) 2019/1381.

[157] Recital 19 Regulation (EU) 2019/1381.

[158] See also the Executive Director's Decision laying down the practical arrangements on pre-submission phase and public consultations, 11.01.2021, available at: https://www.efsa.europa.eu/sites/default/files/corporate_publications/files/210111-PAs-pre-submission-phase-and-public-consultations.pdf). See Vos et al. (2023), p. 57.

[159] Article 32a (1) General Food Law. Article 32a (2) states that "the Authority shall publish general guidance on its website regarding the rules applicable to, and the content required for, applications and notifications, including, where appropriate, general guidance on the design of required studies".

[160] Article 32b General Food Law provides as follows "1. The Authority shall establish and manage a database of studies commissioned or carried out by business operators to support an application or notification in relation to which Union Law contains provision for the Authority to provide a scientific output, including a scientific opinion. 2. For the purposes of paragraph 1, business operators shall, without delay, notify the Authority of the title and the scope of any study commissioned

is made available during the risk assessment stage, rather than only when a final decision is reached. This measure reinforces transparency by requiring applicants to submit *all studies* conducted and collect them in a specific register.[161] The new system therefore seems to provide a more accountable framework for the authorisation procedures. In addition, EFSA verifies the information on the studies carried out and in case of non-compliance the FBOs face procedural consequences such as the non-admissibility of the application or notification.[162] If the application or notification omits previously notified studies, it will not be considered valid.[163] Once the application has been made public according to Articles 38 to 39e, the information about the notified studies should be made public.[164]

Article 32c (1) concerns the application and notification to request the renewal of an authorisation or an approval and the consultation of third parties. In the case of renewals, the substance has already been on the market and to avoid duplication the applicant or the notifier must submit the studies they intend to conduct.[165] EFSA launches a consultation of stakeholders and the public on the intended studies, and after receiving the comments, the Authority shall systematically provide the "renewal pre-submission advice" on the content of the intended renewal application or notification and on the design of the studies.[166] Article 32c (2) provides for the consultation of third parties on the submitted scientific data and studies to identify whether other relevant studies are available on a subject matter implied in an

---

or carried out by them to support an application or a notification, as well as the laboratory or testing facility carrying out that study, and its starting and planned completion dates. 3. For the purposes of paragraph 1, laboratories and other testing facilities located in the Union shall also, without delay, notify the Authority the title and the scope of any study commissioned by business operators and carried out by such laboratories or other testing facilities to support an application or notification, its starting and planned completion dates, as well as the name of the business operator who commissioned such a study. [...]".

[161] See Proposal for a Regulation of the European Parliament and of the Council on the transparency and sustainability of the EU risk assessment in the food chain amending Regulation (EC) No 178/2002, Explanatory memorandum, p. 9.

[162] According to Article 32b General Food Law, "[...] 4. An application or notification shall not be considered valid or admissible where it is supported by studies that have not been previously notified in accordance with paragraph 2 or 3, unless the applicant or notifier provides a valid justification for the non-notification of such studies. Where studies have not been previously notified in accordance with paragraph 2 or 3, and where a valid justification has not been provided, an application or notification may be re-submitted, provided that the applicant or notifier notifies to the Authority those studies, in particular their title and their scope, the laboratory or testing facility carrying them out as well as their starting and planned completion dates".

[163] Article 32b (5) General Food Law.

[164] Article 32b (7) General Food Law.

[165] Proposal for a Regulation of the European Parliament and of the Council on the transparency and sustainability of the EU risk assessment in the food chain amending Regulation (EC) No 178/2002, Explanatory memorandum, p. 9.

[166] See Executive Director's Decision laying down the practical arrangements on pre-submission phase and public consultations, 11.01.2021, p. 2.

application.[167] In addition, Article 32d provides that "in exceptional circumstances of serious controversies or conflicting results" and "without prejudice to the obligations on applicants to demonstrate the safety of a subject matter submitted to a system of authorisation" the Commission may request EFSA to conduct verification studies. Even if the objective is to verify evidence used in the risk assessment process, the studies commissioned may have a wider scope.[168] This "exceptional verification tool" is financed from the Union budget.[169] The Explanatory memorandum has clarified that this procedure forms a risk management tool and there's no overlapping with Article 32 General Food Law on EFSA's capacity to commission studies.[170]

The Transparency Regulation has further modified the rules on transparency and confidentiality contained in Articles 38 and 39 General Food Law to enhance EFSA's independence and address the perceived opacity of the risk assessment. In particular, Article 38 mandates that all scientific data and information supporting requests for authorisation, approvals or scientific opinions should be made public and accessible. EFSA acting "with a high level of transparency", is required to make public, among others, "the scientific outputs, the scientific data, studies" and other information supporting applications, and information on which scientific outputs and opinions are based.[171] The Transparency Regulation, as we will see, takes into

---

[167] Executive Director's Decision laying down the practical arrangements on pre-submission phase and public consultations, 11.01.2021, p. 20, Chapter V on Public consultation on submitted applications, Articles 26–28. The third-party consultation takes place on the basis of the non-confidential version of the application or notification made public by EFSA under Article 38 to 39e.

[168] Article 32d General Food Law.

[169] Recital 25 Regulation (EU) 2019/1381.

[170] See Proposal for a Regulation of the European Parliament and of the Council on the transparency and sustainability of the EU risk assessment in the food chain amending Regulation (EC) No 178/2002, Explanatory memorandum, p. 11.

[171] Article 38, as amended, lists those elements: "(a) agendas, participant lists and minutes of the Management Board, the Advisory forum, the Scientific Committee and the Scientific Panels and their working groups; (b) all its scientific outputs, including the opinions of the Scientific Committee and the Scientific Panels after adoption, minority opinions and results of consultations performed during the risk assessment process always being included; (c) scientific data, studies and other information supporting applications, including supplementary information supplied by the applicants, as well as other scientific data and information supporting requests from the European Parliament, the Commission and the Member Stats for a scientific output, including a scientific opinion, taking into account the protection of confidential information and the protection of personal data in accordance with Article 39 to 39e; (d) the information on which its scientific outputs, including scientific opinions are based, taking into account the protection of confidential information and the protection of personal data in accordance with Articles 39 to 39e; (e) the annual declaration of interests made by the members of the Management Board, the Executive Director and the members of the Advisory Forum, the Scientific Committee and the Scientific Panels, as well as the members of the working groups, and the declarations of interests made in relation to items on the agendas of meetings; (f) its scientific studies in accordance with Articles 32 and 32d; (g) the annual report of its activities; (h) requests from the European Parliament, from the Commission or from a Member State for scientific opinions which have been refused or modified and the justifications for the refusal or modification; (i) a summary of the advice provided to

account existing rules concerning intellectual property rights and provisions related to investments made by innovators ("data exclusivity rules").[172] To ensure transparency, the Authority should implement technical adaptation in order to ensure the efficient processing of requests ("standard data formats", in Article 39f).

Article 39 governs confidentiality requests lodged by the applicants and outlines the procedure, providing a horizontal list of information considered potentially sensitive for commercial interests.[173] Those items (Article 39 (2)) include, for instance: "the manufacturing and production process, including the method and innovative aspect thereof, as well as other industrial specifications", "commercial links between a producer or importer and the applicant or the authorisation holder" or "commercial information concerning market shares or business strategy".[174] Article 39 (4) mandates that the listed information should be disclosed in exceptional circumstances "related to foreseeable health effects or where relevant authorities have identified urgent needs to protect human health, animal health or the environment".[175]

Whether the applicant requests confidentiality, they must provide a verifiable justification and submit a non-confidential version that excludes confidential information (Article 39a).[176] The confidential version contains all information submitted, including the grounds for the confidentiality requests on the different information pieces.[177] The Authority makes its decision on the confidentiality request after a "concrete and individual" examination,[178] having regard to reviewing whether there

---

potential applicants at pre-submission phase pursuant to Article 32a and 32c". Compared with Article 38 before the amendments, the main changes address the elements indicates sub (b), (c) and (d). One of the pivotal adjustments towards an increased transparency is precisely about the disclosure of these information.

[172] Article 38, para. 1a General Food Law.

[173] Recital 32 Regulation (EU) 2019/1381.

[174] Article 39 (2) General Food Law.

[175] Article 39 (4) General Food Law.

[176] Article 39a (1) General Food Law.

[177] Article 39a (2) General Food Law.

[178] It is worth to recall Article 39b (1) General Food Law: "1. The Authority shall: (a) make public the non-confidential version of the application as submitted by the applicant without delay once that application has been considered valid or admissible; (b) proceed, without delay, to a concrete and individual examination of the confidentiality request [...]; (c) inform the applicant in writing of its intention to disclose information and the reasons for that, before the Authority formally takes a decision on the confidentiality requests. If the applicant disagrees with the assessment of the Authority, the applicant may state its views or withdraw its application within 2 weeks of the date on which it was notified of the Authority's position; (d) adopt a reasoned decision on the confidentiality request, taking into account the observation of the applicant, within 10 weeks of the date of receipt of the confidentiality request with respect to application and without delay in the case of supplementary data and information; notify the applicant of its decision and provide information on the right to submit a confirmatory application in accordance with paragraph 2; and inform the Commission and the Member States, where appropriate, of its decision; and (c) make public any additional data and information for which the confidentiality request has not been accepted as justified at the earliest 2 weeks after the notification of its decision to the applicant has taken place pursuant to point (d)".

are urgent reasons concerning human health, animal health, and environment before issuing its scientific output. The information previously qualified as confidential may nevertheless be made public.[179]

Article 39e requires the publicity of the name and the address of the applicant, of the authors of the studies supporting the request and of the participants in the meetings of Scientific Committee and Scientific Panels.[180]

The right of access to documents established in Regulation (EC) No 1049/2001 and in Regulation (EC) No 1367/2001 on access to environmental information shall apply to documents held by the Authority (Article 41). EFSA' Management Board has laid down practical arrangement for implementing the rules on access to documents.[181]

The reform also introduced a "Review clause" in Article 61 relates to the European Commission assessment of EFSA, including an evaluation of Scientific Panels' selection procedure for their transparency and independence. Under Article 61a, Commission experts carry out "fact-finding missions" to evaluate whether laboratories and testing facilities are applying relevant standards in tests and studies submitted to EFSA as part of an application.[182]

[179] Article 39c General Food Law. Article 39d clarifies other obligations with regard to confidentiality. Upon request to the Commission and the Member States, the Authority shall make available "all information in its possession relating to an application or to a request by the European Parliament, by the Commission or by the Member States for a scientific output, including a scientific opinion". In that case, the Commission and the Member States shall take "the necessary measures so that information received by them under Union law for which confidential treatment has been requested is not made public until a decision on the confidentiality request has been taken by the Authority and has become final". Article 39d (3) lays down specific rules in case of an applicant withdraws or has withdrawn an application.

[180] As stated in Recital 34 Regulation (EU) 2019/1381, "It is also necessary to set out specific requirements with respect to the protection and confidentiality of personal data for the purposes of the transparency of the risk assessment process, taking into account Regulation (EU) 2018/1725 and (EU) 2016/679 of the European Parliament and the Council. Accordingly, no personal data should be made publicly available under this Regulation, unless is necessary and proportionate for the purposes of ensuring the transparency, independence and the reliability of the risk assessment process, while preventing conflicting of interests. In particular, for the purpose of ensuring the transparency and to avoid conflicts of interests, it is necessary to publish the names of the participants and observers in certain meetings of the Authority".

[181] See Decision of the Management Board laying down practical arrangements for implementing Regulation (EC) No 1049/2001 and Articles 6 and 7 of Regulation (EC) No 1367/2006 (see Chap. 3 para. 3.3.3. and Chap. 5, para. 5.2). On the Practical arrangements on transparency and confidentiality, see Chap. 5.

[182] Recital 38 Regulation (EU) 2019/1381.

4.5 Dimensions of Transparency in the Obligations Introduced by the New Transparency Rules

After a descriptive presentation of the amendments that the Transparency Regulation operates on the General Food Law, the following paragraphs explore the transparency obligations arising from the different provisions. The analysis uses a key to interpretation that takes into account, when possible, the situation or the context, the goals that are served by transparency in that context, and the potential presence of conflicting interests.[183]

Afterwards, we apply the framework of the dimensions linked with the concept of transparency: democracy, accountability, legitimacy, realisation of individual rights and market efficiency. This allows to understand whether and, if so, to what extent the transparency dimensions are present in the obligations.

4.5.1 Transparency Obligations

The Transparency Regulation introduces a comprehensive *risk communication* framework. As stated before, risk communication is the third component of the risk analysis process. Hence, in the risk communication's obligation, the context regards the risk analysis, whereas the goal concerns exchanging information among the different actors.

Remarkably, the objectives and the principles introduced in Articles 8a and 8b General Food Law seem to broaden the scope of risk communication as it was intended before the new rules. In the new risk communication, the aim to strengthen the citizens' trust is implemented through separate objectives and principles that specify the obligation's content.

Since the risk communication is related to each stage of the risk analysis process, the obligation falls upon both the Commission and EFSA, respectively responsible for communicating risk management and risk assessment decisions. Thus, transparency in risk communication concerns the relationship between the authorities and other interested parties. Among these, the newly introduced Article 8a recalls consumers, food and feed businesses and the academic community. Therefore, widening the involvement of all parties, including those not directly involved in the risk analysis process, is one of the objectives of the new rules. Regarding the presence of conflicting interests, it seems that the objective of fighting against the dissemination of false information in Article 8a takes into account the prevention of practices that may hinder a "transparent" risk communication.[184]

As for Articles 25 and 28, the amendments on the *institutional framework* of the Authority follow the purpose of increasing consumers' trust through measures

---

[183] See Buijze (2013), p. 10.

[184] On the issue of false information in the food sector, see González Vaqué (2015), pp. 411–420.

designed to ensure EFSA's independence. In this section, the Regulation frames transparency obligations as rules amending the governance. The context regards the Management Board's composition and the best independent experts' engagement to work in the Scientific Panels and Committee. The obligation falls under the Authority's remit, and it seems not possible to identify conflicting interests.

As regards the *adjustments to the risk assessment*, we must distinguish the different obligations introduced by the Transparency Regulation with the aim of identifying common features and peculiarities. The situation refers to the application or notification in relation to which Union law sets provision for EFSA to provide a scientific output.

Under Article 32a applicants may request a "pre-submission advice" in the phase prior to the submission. That is useful for offering information and clarifying the businesses' duties to facilitate their compliance. As already mentioned, to prevent conflicting interests, the staff responsible for the preliminary advisory part is not involved in the scientific assessment of the application or notification, so the results are not affected by the pre-submission advice.

Article 32b requires FBOs to disclose the studies they have commissioned when preparing an application or a notification. It forces businesses and laboratories to be transparent as they must provide information regarding those studies. To facilitate this, EFSA should establish and manage a database of studies as a means to increase guarantees ("database of study notifications" or "database").[185] The purpose of this obligation is to prevent businesses from withholding unfavourable studies. Businesses can have a conflicting interest in disclosing all the studies because they can reveal unwanted information that may affect the application's outcome.

Nevertheless, the purpose of the Transparency Regulation is precisely to ensure transparency of *all studies* and information provided by the industries during the risk assessment process. As a consequence, and in combination with the rules on transparency laid down in Article 38, the Authority makes public the studies when receiving the corresponding application or notification once they are considered valid or admissible. This means that the public can access the information contained in the studies and express their opinions. At the same time, the Authority takes into account other conflicting interests detained by the industries regarding the requests for confidentiality, which is guaranteed in duly justified circumstances.

Article 32c imposes a similar obligation on businesses seeking to renew the authorisation or the approval of a certain substance already on the market, focusing on renewals. The goal is to avoid repetitions and to check the new studies. In this case, and in all cases of application for an authorisation/approval, the industries fall under the obligation to submit the studies, whereas the Authority shall launch a consultation of stakeholders and the public. Therefore, stakeholders and the public are actively involved in the renewal procedures, and they can express their opinions on the studies. Even in this context, the Authority shall consider the industries'

[185] See Executive Director's Decision laying down the practical arrangements on pre-submission phase and public consultations, 11.01.2021, Article 17.

conflicting interests on confidentiality. Thus, the consultation occurs on the non-confidential version of the request.

As for Article 32d, the obligation to commission scientific studies falls under the EFSA's remit following a Commission request. Situations such as controversies or inconsistent results demand greater clarity in the studies. The goal is to bring different studies to the Authority's attention to solve the mentioned inconsistencies. At the same time, this operation should not compromise the businesses' interests to demonstrate a subject matter's safety submitted for the authorisation.

Article 38 lays down a set of transparency obligations on EFSA. Namely, the Authority should disclose a series of documents and studies in the context of its operations. This Article is linked with all the previous because it reshapes the obligations concerning EFSA's activities and its duty to *be transparent*. More specifically, Article 38 appears as an example of *active models* of transparency. As we mentioned in Chap. 3, in those models an institution provides public access to information without a prior request. The goal is to allow the participation of stakeholders and the public in the Authority's work. In this context, the definition of "stakeholder" includes consumers, environmental/health NGOs and advocacy groups, farmers and primary producers, business and food industry, distributors' associations, and academics.[186]

In its engagement with those stakeholders, the Authority should bridge the transparency deficit resulting from the Fitness Check. Consequently, the public can be involved in EFSA's decision-making process. However, the extensive obligation set up in Article 38 cannot compromise other conflicting interests, such as intellectual property rights and data supporting applications for authorisations. Therefore, the new Article 39 sets some *derogations* to the transparency obligations which count as mitigation of transparency and not as transparency obligations. Hence, the procedure established in Articles 39-39d corresponds to the different purposes of examining the businesses' confidentiality requests.

A different regime applies to the request for *access* to documents. Article 41 deals with applications for access to documents held by the Authority. The aim is to give access according to the relevant rules set out in the corresponding Regulations, while balancing any interests conflicting with the disclosure pursuant to the law.[187]

4.5.2 Application of the Transparency Dimensions

After describing the General Food Law's amendments introduced by the Transparency Regulation and reflecting widely on the new transparency obligations, we will apply the transparency dimensions framework.

[186] See EFSA (2016), p.22. See also Vos et al. (2023), p. 55, Chearnaigh (2021), p. 705.

[187] See Article 4 Regulation (EC) no 1049/2001.

The *democratic dimension* of transparency concerns the visibility of the public authorities' power. The transparency of authorities' actions tends to ensure the acts and measures taken towards citizens. Transparency and the development of participatory organisational models enhance democracy. Hence, it appears that the democratic dimension is present in several obligations introduced by the Transparency Regulation.

For instance, risk managers and risk assessors must provide information to the public in risk communication. Information is a way to improve democracy through transparency and disclosure. Stakeholders and citizens are better involved in the institutions' decisions; therefore, they participate in the risk analysis process, increasing its democratic value.

Regarding the amendments on EFSA governance, the participation of representatives of Member States, of the European Parliament and the Commission, civil society and industry organisations in the Management Board can be seen as a way to "democratize" the Authority. In this view, the aim of increasing participation comes directly from changes in the institutional composition. As recalled in Chap. 2, the White Paper on European Governance links transparency to confidence in experts' advice. Even the changes in the appointment mechanism of the experts and the verification of their independence criteria could be viewed as a means to enhance democracy.

The democratic dimension occurs in other transparency obligations. For instance, Article 32c requires an open participation of stakeholders and the public in the decisions on authorisations. In this respect, the Authority shall launch a consultation and involve the public in the renewal procedures so that they can express their opinions. This aspect improves the democratic dimension of transparency.

Furthermore, Article 38 concerning transparency of the Authority is undoubtedly a means for enhancing democracy. EFSA should disclose information, such as scientific outputs, scientific data, studies and other information supporting applications. In this way, the Authority improves democracy through the information's visibility.

The second dimension to consider is *legitimacy*. By legitimacy, it is intended the public acceptance of the decision-making procedure. It is a way to enhance citizens' trust in the public authorities' actions. Legitimacy also reduces the distance between the public and the administration. Within the new provisions of the General Food Law, this dimension seems to occur mainly in the amendments on EFSA's structure and in some transparency obligations. An efficient risk communication, as outlined in the new rules, enhances legitimacy. Citizens can be aware of the risk analysis process and understand its outcome, thereby increasing public acceptance of the procedure.

The reform strengthens the Member State's involvement in the EFSA's governance and improves the scientific experts' independence. The increased Member States' engagement in the risk assessment phase can enhance legitimacy because they can feel represented in the Authority's governance and take part in the decisions. Consequently, citizens' trust in the Authority's organisation and functioning may improve. It has been noted that transparency in policy making allows people to see how experts are selected, and this generally promotes legitimacy. Thereby, it can be argued that the amendments of the appointment mechanism in order to increase the experts' independence moves in that direction too. Moreover, granting public access to the studies submitted by the industries and enabling stakeholders to express their opinions on them enhances legitimacy. It is evident that involving the stakeholders and the public increases the degree of transparency in the whole process. Thus, the obligations set out in Article 32b and 32c requiring the submission of the studies by the businesses aim to improve legitimacy, even though the final decision on authorisation rests with the public authority.

Similarly, on the side of EFSA, Article 38 could be interpreted as a tool that enhances legitimacy by disclosing information. The information, especially that on which scientific outputs and opinions are based can improve legitimacy. That is because showing the information which had an impact in the decision-making process involves citizens more closely. Therefore, they may accept or even agree on the authorities' determinations.

A different transparency dimension concerns *accountability*. Accountability is the duty of an individual or organisation to answer about how they have conducted their affairs.[188] In sum, it means to be accountable for one's actions, and it is a concept mostly used to qualify an actor's performance. Transparency in relation to accountability can help alleviate the information asymmetry and improve governance quality.

By analysing the transparency obligations, it can be realised that they reveal the accountability dimension.

It seems pretty clear that the new obligations on risk communication enhance accountability. Providing information on the risk analysis process makes the Commission and EFSA accountable for their decisions. This can be confirmed by considering Recital 4 of the Transparency Regulation. It asserts that risk communication should foster a participatory and open dialogue to ensure that the risk analysis process is based on, among other values, *accountability*.

Furthermore, the Member State's engagement in EFSA and the experts' new appointment mechanisms are meant to ensure that the Authority is more accountable to the Union citizens.

We can note the presence of the accountability dimension also in the new obligations on the risk assessment. In this context, the duty to disclose the studies commissioned makes the businesses more accountable. Moreover, the obligation to disclose the information on the commissioned studies serves to verify that the Authority is accountable and does not rely only on the studies commissioned by the industry when providing the scientific output. Therefore, the accountability dimension can also be found in Articles 32b, 32c and 38. They all pursue the scope to reduce information asymmetries. As regards Article 32d, we can qualify the obligation falling on EFSA to commission other verification studies as a means to improve accountability. In this situation, the Commission ascertains the presence of

[188] See Chap. 2.

conflicting results and aims to solve it, acquiring more evidence. Hence, on the back of the new scientific evidence provided, the Commission can decide and ensure greater accountability to the Union citizens.

Transparency can also *realise individual rights*. In this context, it has a strong link with individual expectations. From a rights-based perspective, transparency is a tool for promoting people's rights. The question is whether the reform considers the individual rights within the transparency obligations, and particularly the right of access to documents.

The right of access to documents is enshrined in Article 41. Even before the amendments, the European legislative framework on access to documents—encompassing Regulation (EC) No 1049/2001 and Regulation (EC) No 1367/2006 in case of environmental information—applied to the documents held by the Authority. Now, Article 41 explicitly refers to the abovementioned Regulations. There is no doubt that the individual request to access the documents held by EFSA should follow those disciplines.

Another individual right can be identified in the confidentiality request that a business may ask the Authority according to Articles 39-39d. We have already expressed that these Articles set some derogations to transparency obligations rather than providing them. Consequently, they constitute a means to balance the rights of access in the transparency dimensions' framework with the right of confidentiality.

As argued, transparency can increase the level of *market efficiency* and economic performance. The availability of information facilitates decision making and affects the resource management. Therefore, transparency contributes to the proper functioning of the market and the contractual relationships.

From the wide analysis of the transparency obligations, we cannot conclude that specific ones aim to improve market efficiency. Rather, it can be argued that the availability of information and the measures to intensify transparency which derive from the overall Regulation may increase the market efficiency. The consequences on the market effectiveness could be verified only after the implementation of the new rules. Nevertheless, from a comprehensive evaluation of the transparency obligations, we could argue that, besides the scope to increase transparency, the Transparency Regulation has considered the context in which businesses operate. For instance, Article 32a provides that the Authority should give pre-submission advice upon applicants' requests before the application or notification is formally submitted. The fact that actors such as small and medium-sized enterprises do not always have a clear understanding of the application's requirements justifies this measure. Therefore, the advice facilitates compliance by the applicants and may speed up the submission process. More generally, this simplification may affect the overall authorisation procedure, and thus also the potential placing on the market.

From the same perspective, the risk assessment's adjustments seem to consider also the market context. One could argue that the new rules increase the burdens on the applicants and have a negative impact on the effectiveness. By contrast, the improvements in terms of transparency may have the effect of simplifying the overall performance of the authorisation or renewal procedure. In this regard, the Proposal explicitly clarified that the notification of commissioned studies creates a minimal burden and takes place at the pre-submission stage. At the same time, the risk assessment will run in parallel to the consultation on the submitted studies.[189] Thus, the impact in terms of timing will be minimal. Additionally, the notification of planned studies in the renewals context represents a relatively small burden for the applicant.[190] It can also be considered proportionate given that at an early stage of the process, the applicant can get useful advice on the content of its intentional application as a result of the consultation.[191]

Given this context, it is possible to conclude that the reform's impact may improve the market efficiency in the authorisation procedure. Clearly, this statement should be verified after the implementation of the new rules over a certain observation period.

In conclusion, all the transparency dimensions identified in the first stage of this book seem to occur in the transparency obligations set out by the Transparency Regulation. Notably, not every transparency obligation presents all the dimensions described. It is unlikely that a single transparency obligation would cover all the dimensions identified. It may address only some of them or even other dimensions beyond the one outlined in this book. It is not essential to *quantify* how many dimensions are present in each obligation. On the contrary, the focus lies in assessing whether and to what extent transparency is detected in the legislation under scrutiny. Accordingly, classifying the dimensions of transparency serves to identify the various forms in which transparency can be manifested. Since the new Regulation is specifically designed to introduce a set of transparency obligations in the General Food Law, it can be concluded that all the previously identified dimensions are, to some extent, reflected in the new rules.

Nevertheless, we can state that the reform takes into account mostly the dimensions of democracy, legitimacy and accountability. Undoubtedly, the new rules' scope focuses on the consumers' perception concerning risk assessment. They aim to increase the consumers' trust in the risk analysis process. To this extent, they also seek to improve the Authority's independence. Consequently, the majority of the amendments focus on aspects that emphasize the position of EFSA. In this vein, transparency arises primarily in the procedures and the Authority's role. Thus, there is consistency between the aim of the Regulation and the content of the obligations.

---

[189] Proposal for a Regulation of the European Parliament and of the Council on the transparency and sustainability of the EU risk assessment in the food chain amending Regulation (EC) No 178/2002, Explanatory memorandum, p. 9.

[190] Proposal for a Regulation of the European Parliament and of the Council on the transparency and sustainability of the EU risk assessment in the food chain amending Regulation (EC) No 178/2002, Explanatory memorandum, p. 9.

[191] Proposal for a Regulation of the European Parliament and of the Council on the transparency and sustainability of the EU risk assessment in the food chain amending Regulation (EC) No 178/2002, Explanatory memorandum, p. 9.

4.6 Dimensions of Transparency in the "Unmodified" Provisions of the General Food Law

After examining the Transparency Regulation and the obligations respectively introduced, we move back to the General Food Law. The analysis addresses whether there are transparency obligations in the General Food Law provisions not amended by the Regulation (EU) 2019/1381. The latter explicitly modifies the General Food Law with the scope of improving transparency. Therefore, the key to interpreting its provisions was assessing how this aim resulted at a more concrete level within the transparency obligations.

Instead, while analysing the General Food Law in the Sections and Articles not touched by the reform, we will focus on the definition of transparency. When defining transparency, we chose the angle of the *availability of information* and the *exchange of information between different actors*. Information is a crucial concept, and we intend it as an *interaction* among subjects. Once established that the reference to information may lead to recognising a transparency obligation, we could consider the situation or context, the goal and potential conflicting interests using the same aforementioned approach.

The focus on information allows us to examine how the transparency obligations incumbent on public authorities and private actors in the General Food Law are enshrined.

4.6.1 Transparency Obligations

Besides the reference and the specifications on the risk communication in Article 8, the general principles are not affected by the reform.

Significantly, the Transparency Regulation has not amended the Section called "Principles of Transparency" encompassing Articles 9 and 10 General Food Law. However, we should recall these principles, given that they directly express the transparency concept in the General Food Law.

Article 9 regards the context of the preparation, evaluation and revision of the food law. The public consultation principle aims to solve social control issues in food safety policies and seek consumers consensus on these policies. As we noted with regard to the target of the transparency obligation, in a democracy, the relationship between citizens and institutions is grounded on reasons of control or trust.[192] In Article 9, transparency is based on consensus-seeking policies, and information is necessary to sustain consumers' trust.

The information and participation model of rulemaking has a greater capacity to bring problems to light outside the traditional channels of expert committees.[193] It

[192] See Chap. 2, para. 2.4.

[193] Viti (2003), p. 225.

also dispels suspicions of strong links between experts who determine technical rules and food industry lobbies.[194] In sum, the method of information and participation shifts into a dialectical system in which it is possible to compare different opinions.

Besides consultation, Article 10 represents an institutional approach to civil society involvement in food safety policies. In this regard, public authorities have the duty to *inform on the risks*. Therefore, the principle of transparency must be applied both before adopting regulatory measures and after the possible emergence of situations requiring decisions that may affect consumers' generality.

Even after adopting the new rules on transparency, these Articles remain essential guides to interpreting transparency in the General Food Law.

Moving to the analysis of the *general requirements of food law*, we must consider Article 14 General Food Law.[195] At a closer look, Article 14 General Food Law contains a transparency obligation in the third paragraph. In particular, determining whether a food is unsafe, Article 14 (3) General Food Law recalls the *information* provided to the consumer, including information on the label, or other information on the use of specific products and on possible adverse effects.[196] In this vein, the food business operator is obliged to give information to the consumer with the aim of being transparent on the use and impacts of the product. It has been argued that food safety is not considered as a sort of "abstract and objective paradigm" in which a product is safe by itself.[197]

On the contrary, a product can be considered safe also by a "subjective and conscious approach to the use of the product".[198] Accordingly, we should not consider the consumer using food only as a weak actor to be protected, as the consumer should play an active role in the food safety system.[199] In this view, the transparency obligation is a duty on the food businesses that should provide fair and exhaustive information. In addition, the transparency obligation in Article 14 should also take into account the consumer's role in reading the information to avoid the adverse health effects. In this regard, one can argue that consumers should be clear-eyed about the food label's nature and treat it like a legal document offered for their

[194] Viti (2003), p. 225.

[195] In the sense of a connection between the principle of transparency and Article 14 GFL, see also Bolognini (2013) p.163.

[196] For convenience, we report the text of Article 14(3) General Food Law: "In determining whether any food is unsafe, regard shall be had (a) to the normal conditions of the use of the food by the consumer and at each stage of production, processing and distribution, and (b) to the information provided to the consumer, including information on the label, or other information generally available to the consumer concerning the avoidance of specific adverse health effects from a particular food or category of foods".

[197] Albisinni (2015), p. 111.

[198] Albisinni (2015), p. 111.

[199] Albisinni (2015), p. 111

assent.[200] We can conclude that whereas Article 14 imposes a transparency obligation toward food consumers concerning information, it also requires their "active role". This role does not necessarily take the form of a conflicting interest. Instead, the consumer's role as a receptor of the information completes the transparency obligation.

Regarding the other provisions not amended by the Transparency Regulation, the same Section on the *general requirements of food law* presents the rules on presentation, traceability and responsibility of food business operators. Articles 16, 17, 18 and 19 General Food Law seem to have a connection with transparency and place transparency obligations on food business operators.

Article 16 requires that the *information* carried through labelling, advertising and presentation—including the shape, appearance or packaging—shall not mislead the consumer. We must immediately connect this Article with the principle of protecting consumers' interests laid down in Article 8 General Food Law.[201] As said before, the consumers' interest principle is one of the general principles that shall form a "general framework of horizontal nature".[202] More generally, Article 16 expresses the consumer's demands in healthy and safe food that dominate the entire General Food Law.[203] It is certainly important to note that even in this case, the *presentation* of food seems to become an obligation on the food business operator's side to be transparent towards the consumer. To this aim, industries act in different contexts concerning labelling and advertisement. On the contrary, the consumer has an interest in receiving non-misleading information. In the obligation under Article 16, similarly as in Article 14 (3), the consumer plays an active role and can make an informed choice.

As already mentioned, Article 17 (1) General Food Law places on the food business operator the primary responsibility for ensuring compliance with food law and in particular food safety. In this system, businesses must ascertain whether food on the market is *unsafe*, rather than providing safe food or justifying food safety. Companies have the duty to *prevent* food safety risk through compliance with the

---

[200] Heinzerling (2015), p. 23. According to the author, "[the food label] has been drafted by strangers, mostly for strangers' benefit. Every word, every number, even every font, has been parsed and negotiated and approved by strangers' lawyers. It is not a neutral document, and it is not authorless, even if it is anonymous. The ancient impulse of *caveat emptor* should activate, not recede, upon seeing a label on the food one is about to buy".

[201] Di Lauro (2003), p. 260.

[202] See, para. 4.2.2.3.

[203] Di Lauro (2003), p. 260.

hygiene regulations,[204] to be *prepared* in the form of traceability,[205] and the duty to *respond* in case of recall.[206]

It is worth focusing on the implication on businesses, leaving aside for a moment the obligations that Article 17 (2) places on Member States, among which there are the adoption of a system of official controls and a framework of rules and penalties for food law infringements. A closer look at Article 17 (1) allows us to catch the transparency profile in this norm. The goal is to ensure food safety through the industries' responsibilities covering all food chain sectors.

In order to clarify the content of the operators' responsibility, the Court held in the *Lidl Italia* case that, notwithstanding the retailer does not have immediate control on the infringement, both the producer and the retailer can be held liable for a non-compliance with food legislation.[207] Similarly, the Court declared that national law could impose sanctions on "a food business operator which is active only at the distribution stage for placing a contaminated food product on the market".[208]

Thus, the General Food Law has introduced a system of responsibility throughout the food chain, both for information to consumers and ensuring food safety compliance.[209] In this context, it has been noted that in complying with Article 17 (1), food business operators must be *transparent in the internal organization.*[210] For instance, businesses can ensure transparency through the adoption of the HACCP or traceability procedures.[211]

Conversely, Article 17 (2) seems to make a reference on transparency in requiring Member States to ensure *public communication* on safety and risk on food. In this case, transparency takes the shape of an information flow toward citizens.

---

[204] See Regulation (EC) No 852/2004 of the European Parliament and of the Council of 29 April 2004 on the hygiene of foodstuffs OJ L 139, 30.4.2004; Regulation (EC) No 853/2004 of the European Parliament and of the Council of 29 April 2004 laying down specific hygiene rules for food of animal origin, OJ L 139, 30.4.2004. See also Spirovska Vaskoska et al. (2020), pp. 304 ff.

[205] See Article 18 General Food Law.

[206] See Article 19 General Food Law.

[207] Judgment of the Court (Second Chamber) of 23 November 2006, *Lidl Italia Srl v Comune di Arcole (VR),* Case C315/05, ECLI:EU:C:2006:736. See Bartl (2015), p. 87; Klaus and Meyer (2008), p. 6; Santini (2013), p. 257. As expressed by Santini, "[the Lidl case] drew attention on something that has clearly been viewed as a lacuna in European food law needing to be filled: the recent Regulation (EU) No 1169/2011 on the provision of food information to consumers in facts provides that the operator responsible for the food information shall be the one under whose name or business name the food is marketed (or, if that operator is not established in the Union, the importer into the Union market)". See also Rubino (2013), pp. 303 ff.

[208] Bartl (2015), p. 87.

[209] Bartl (2015), p. 88. The author recalls Article 8(1) Regulation (EU)1169/2011 on food information to consumer ("FIC") as a possible solution for the challenge on food business operators to find a balance "between undertaking the necessary level of due diligence and the high costs related to sample testing and product checks".

[210] Albisinni (2015), p. 112.

[211] Albisinni (2015), p. 112.

Regarding traceability, Article 18 General Food Law requires the industries to put in place a system and procedures that allow *information* on their product to be made available to the competent authorities on demand.

More in detail, the traceability system aims "to enable food safety problems to be identified at the source, and across the food chain".[212] Thus, the purpose is to quickly identify the source of food safety problems and conduct potential recalls to take the product from the market. To this extent, businesses must register precisely where their food material originated and where it went.[213] The Conclusions of the Standing Committee on the food chain and animal health related to the Guidance on the implementation of Articles 11, 12, 14, 17, 18, 19 and 20 express that traceability records help to "facilitate targeted withdrawal and recall of food, thereby avoiding unnecessary disruption of *trade*", "enable consumers to be provided with accurate information concerning implicated products, thereby helping to maintain consumer confidence", and "facilitate the risk assessment by control authorities".[214] Apart from its role in food safety, traceability also helps to ensure fair trade relationships among operators and the reliability of *information supplied to consumers* in terms of "substantiating claims" made by producers.[215] This way, it aims to reconcile the interests of food producers in the proper functioning of the market with those of consumers in healthy and safe food.

Specific requirements outlined in Article 18 General Food Law mandate that food business operators must be able to trace their inputs and outputs "*one step back and one step forward*". To this end, FBOs must identify "any person from whom they have been supplied with a food, a feed, a food-producing animal, or any substance intended to be, or expected to be incorporated into a food or feed".[216] They must also be able to identify the other businesses to which their products have been supplied.[217] Companies must implement systems and procedures to comply with the authorities' requests for information. In addition to this kind of information, businesses should provide other documentation or *information* on traceability through a system of food labelling and identification.[218]

---

[212] Spirovska Vaskoska et al. (2020), p. 315.

[213] Spirovska Vaskoska et al. (2020), p. 315. Article 3(15) defines traceability as "the ability to trace and follow a food, food-producing animal or substance intended to be, or expected to be incorporated into a food or feed, through all stages of production, processing and distribution".

[214] See Guidance on the implementation of Articles 11, 12, 14, 17, 18, 19 and 20 of Regulation (EC) No 178/2002 on General Food Law, para III.1 p. 16. Emphasis added.

[215] Guidance on the implementation of Articles 11, 12, 14, 17, 18, 19 and 20 of Regulation (EC) No 178/2002 on General Food Law, para III.1, p. 16.

[216] Article 18 (2) General Food Law.

[217] Article 18 (3) General Food Law.

[218] Article 18 (4) provides that "food or feed which is placed on the market or is likely to be placed on the market in the Community shall be adequately labelled or identified to facilitate its traceability, through relevant documentation or information in accordance with the relevant requirements of more specific provisions". More specific legislation is provided for certain sectors, e.g. Commission Implementing Regulation (EU) No 931/2011 of 19 September 2011 on the traceability requirements set by Regulation (EC) No 178/2002 of the European Parliament and of the

From a closer look, it appears that in the traceability requirements there are transparency obligations. These consist of providing information in business relationships and to authorities upon request. Moreover, duties linked to the system of labelling can be either seen as transparency obligations.[219]

Given this context, some conflicting interests can arise. Businesses at the first stage of the food chain where a problem originated could see as a disadvantage the possibility of using traceability to establish liability of the business that caused a food safety problem.[220]

It has been recognised that the relationship between food safety and transparency shows how companies try to react to food safety concerns.[221] Consumers and other stakeholders are increasingly concerned about the ongoing sequence of food scandals and incidents.[222] Given the continuous need for transparent information on the quality of the entire food chain, businesses can extend their assortments by products and brands associated with safety and quality and formally register their performance.[223] Moreover, companies adopt a technical system to communicate and improve their quality performance and claim their integrity about food safety and quality in communication with stakeholders.[224] Due to the complexity of the food chains, transparency is a difficult task. Thereby, industries often need to enhance transparency through innovations in technologies.[225]

Article 19 establishes an obligation on the FBOs to withdraw from the market food not compliant with food safety requirements. In case the product has reached the consumers, the businesses have to *inform* the consumers of the reason for the withdrawal and recall the product. The procedures to withdraw from the market shall be initiated by the operator responsible for retail or distribution activities. The

Council for food of animal origin, OJ L 242, 20.9.2011; Commission Implementing Regulation (EU) No 208/2013 of 11 March 2013 on traceability requirements for sprouts and seeds intended for the production of sprouts, OJ L 68, 12.3.2013, Regulation (EC) No 1830/2003 of the European Parliament and of the Council of 22 September 2003 concerning the traceability and labelling of genetically modified organisms and the traceability of food and feed products produced from genetically modified organisms and amending Directive 2001/18/EC, OJ L 268, 18.10.2003. See Spirovska Vaskoska et al. (2020), pp. 315 ff.

[219] See also Pisciotta (2018), pp. 36 ff. According to the author, Articles 16 and 18 General Food Law are "inspired" by the principle of transparency.

[220] Spirovska Vaskoska et al. (2020), p. 316.

[221] Beulens et al. (2005), p. 481; Astill et al. (2019), p. 241.

[222] Beulens et al. (2005), p. 481.

[223] Beulens et al. (2005), p. 482. Industries can create also their own legal frameworks on the basis of contractual law instruments. This "self-regulation" can be an answer to food safety problems with the creation of quality standard. See Nouwt et al. (2003), p. 3.

[224] Beulens et al. (2005) p. 482. On the administrative burdens related to traceability in the European food industry, see Bremmers et al. (2008), p. 49 ff.

[225] Astill et al. (2019), p. 241 ff. analyze the technologies provided by the Internet of Thing (IoT) which have the potential to increase food production transparency. IoT is a system consisting in sensors and advices that collect and transfer data and require a range of enabling technologies.

latter should also inform other actors such as producers, processors, manufacturers and authorities in order to trace the food.

The obligation to withdraw, recall or notify unsafe food arises when the food is unsafe under Article 14 General Food Law.[226] Article 19, similarly to the obligation laid down in Article 17, imposes on businesses to be transparent to consumers. This time, the information is given after the sale, and it should be accurate and effective on the reasons for withdrawal.[227] Thus, among the transparency obligations arising in the section on the general requirements, we can also include Article 19 on responsibilities.

With regard to the chapter on European Food Safety Authority, from an analysis of the legal text, it is possible to note that the Transparency Regulation has modified most of the provisions regarding the exchange of information and transparency. We must therefore refer to the earlier comments on the changes introduced on EFSA. It may be interesting to observe that the new Regulation has not changed the provisions concerning the Rapid Alert System and crisis management in the section on the Authority nor in the dedicated Section (Articles 50–60).

Therefore, it can be useful to consider the RASFF system and the emergencies in order to verify whether there are any transparency obligations.

As mentioned in the descriptive part of the General Food Law, the Rapid Alert System is an instrument to communicate food and feed risks.[228] It is created to allow authorities to be rapidly informed on the risks and take any necessary action. According to EU law provisions, a food alert is an international system for urgent notification of food or feed risks connecting the European Commission and Member States, with the possible inclusion of other countries.[229] Under the risk analysis principle, the decision to communicate a risk through the rapid alert system is a risk management decision, which can only be taken after a risk assessment.[230] Moreover, consultation with interested parties and evaluating other legitimate factors are a part of the process.[231] Consequently, a food alert results from a regulated process.

[226] Guidance on the implementation of Articles 11, 12, 14, 17, 18, 19 and 20 of Regulation (EC) No 178/2002 on General Food Law, para IV.3 p. 23.

[227] Guidance on the implementation of Articles 11, 12, 14, 17, 18, 19 and 20 of Regulation (EC) No 178/2002 on General Food Law, para IV.3, p. 25.

[228] See para 4.2.6. The RASFF was introduced in the EU as a tool for the notification of risk on human health and as a "reaction of the episode of "food terrorism"" concerning the international contamination with mercury of a cargo of orange from Israel. The Rapid Alert System was created by a proposal for a Council Decision (COM/79/725 final) and defined as a gentlemen's agreement. Then, an amended proposal and a Council Decision occurred in the 1984 (84/133/EC) created "a system for the rapid exchange of information on dangers arising from the use of consumer products", including both food and non-food. Later, the Rapid Alert System was subject of Decision 89/45/EEC and regulated by Directive 92/59/EEC on general product safety. Finally, the General Food Law created the RASFF. See Fuentes (2019), p. 279.

[229] Fuentes (2019), p. 278.

[230] Fuentes (2019), p. 278.

[231] Fuentes (2019), p. 278.

More in detail, the General Food Law lies down in Article 50 the legal basis for the RASFF system, which creates a network for the notification of a direct or indirect risk to human health deriving from food or feed.[232] The network of authorities links all Member States, the European Food Safety Authority and the European Commission which manages the system. From the beginning, the system has been designed as open to applicant countries, third countries or international organisations on the basis of agreements.[233] All the members must designate a contact point which shall be a member of the network and facilitate the communication.[234]

The use of the system is mandatory in case of serious risk to human health[235] when a Member State adopts any measure aimed at restricting the sale of food or at withdrawing and recalling in order to protect human health, or in case of border rejection related to risk to human health.[236] In all these cases, the notification is based on a system of *information* accompanied by the reasons for the action taken by the Member State authorities in which the notification was issued, and it is followed by supplementary information.[237] Commission Regulation (EU) No 16/2011 extended the use of the RASFF to another kind of communication called "information" which refer to situations where the health risk is less serious and does not require immediate action.[238] The discipline is under Commission Implementing Regulation (EU) 2019/1715, which maintains the three categories of RASFF notification,[239] called "alert", "information" and "border rejection".[240]

Interestingly, in the rules on RASFF, there is also a mention of the principle of transparency as laid down in Article 10 General Food Law. Namely, Article 52 states that information communicated by RASFF shall be available to the general public "in accordance with the information principle provided for in Article 10"[241] to the extent that this *public information* is necessary to protect human health. Among the information available to the members of the network, the ones concerning

---

[232] Article 50 (1) General Food Law.

[233] Article 50 (6) General Food Law. At the moment, the European Commission, the European Food Safety Authority, all Member States, the European Free Trade Association (EFTA) are members of the system (see https://ec.europa.eu/food/safety/rasff/members_en).

[234] Article 50 (1) General Food Law.

[235] Article 50 (2) General Food Law.

[236] Article 50 (3) General Food Law. Article 50 (3) includes also the case in which the notification concerns "any recommendation or agreement with professional operators which is aimed, on voluntary or obligatory basis, at preventing, limiting or imposing specific conditions on the placing on the market or the eventual use of food or feed on account of a serious risk to human health requiring rapid action". The principle expressed in the *Karl Berger* case (C-636/11) is applicable to the RASFF, see para. 4.2.2.4.

[237] Article 50 (3) General Food Law.

[238] Fuentes (2019), p. 281; Fuentes (2017), p. 122.

[239] Commission Implementing Regulation (EU) 2019/1715 of 30 September 2019 laying down rules for the functioning of the information management system for official controls and its system components (the IMSOC Regulation), OJ L 261, 14.10.2019.

[240] For some examples and statistical reviews, see Parisi et al. (2016), p. 19.

[241] Article 52 General Food Law.

product identification, the nature of the risk and the measures taken must be disclosed. Nevertheless, information covered by professional secrecy must not be revealed except in duly justified cases.

Consequently, it could be argued that even in the case of RASFF—where the focus is on the communication within a network—transparency emerges. The availability of information to disclose to the public consists indeed in a "transparency obligation". In this context, a conflicting interest is the protection of professional secrecy. Business may suffer from reputation damages and financial losses in cases where the information is divulged that their products have been declared to be at risk. In the *Malagutti case,*[242] a company exporting fruit from France to the UK and the Netherlands has been named in an alert notification as a supplier of apples containing a higher content of pesticide residues than the maximum permitted level. As Malagutti's liability was excluded by the French authorities, the company brought an action before the Court of First Instance (CFI) suing the European Commission for extra-contractual liability. The Court, in considering not admissible the action, specified that the cooperation between the Commission and the national authorities of the Member States "cannot cause the Community to incur liability towards individuals since the alert system is ultimately triggered on the initiative of, and according to an analysis made by, the national authorities alone".[243]

However, it has been noted that reputation can be seriously damaged because the impact of the RASFF is more about the publication than the existence of the RASFF itself.[244]

Another provision on crisis management expressly recalls the principle of transparency. That is to say Article 55 General Food Law on the general plan for crisis management. The Commission shall draw up, in close cooperation with the Authority and the Member States, a general plan for crisis management in the field

---

[242] Judgment of the Court of First Instance (Second Chamber) of 10 March 2004, *Malagutti-Vezinhet SA v Commission of the European Communities*, Case T-177/02, ECLI:EU:T:2004:72.

[243] See *Malagutti-Vezinhet SA v Commission,* paragraph 27. The Court dismissed the claim for damage because it upheld that under the RASFF only the national authorities are responsible for establishing whether there is a serious and immediate risk to the health and safety of consumers: "The applicant should therefore have referred the matter to the national court having jurisdiction. The question of compensation by a national agency for damage caused to private individuals by national agencies, either by reason of an infringement of Community law or by an act or omission contrary to national law, must be determined by the national courts (Case 101/78 Granaria [1979] ECR 623). The applicant has by no means demonstrated that a claim for compensation brought before the national courts of any of the States involved would not have enabled it to obtain fair compensation for the damage at issue" (paragraph 28). See also Santini (2014).

[244] Fuentes (2019), p. 282 ff. Although the Regulation requires respect to the principle of confidentiality, the fact that so many different independent authorities participate in the network makes confidentiality difficult to preserve. The reputation can be indeed seriously damaged without disclosing any professional secrecy. The RASFF in practical terms is not only an internal means of communication between authorities because it can have devastating external effects. That is because the nature of information transmitted through it, which regards food-related health problems. Furthermore, the information is transmitted instantaneously to a great number of authorities in different countries. See also Fuentes (2010), p. 204 ff.

of food and feed safety. The general plan was adopted by Decision (EU) 300/2019.[245] The General Food Law defined the content of the general plan in vague terms. Article 55 (2) requires that the plan shall specify the types of situations involving risk to human health which are not likely to be prevented, eliminated or reduced by the other provisions.

Health risks cannot be addressed by simply applying Articles 53 and 54 on the adoption of emergency measures. Thus, the general plan seems to take on the role of an *extrema ratio* tool. The plan shall specify the practical procedures necessary to manage a crisis. Among these, it encompasses the *principle of transparency* and a communication strategy. Accordingly, the plan entails gathering scientific information related to the crisis and a communication strategy, including preliminary contact with stakeholders.[246]

The reference to transparency follows the principle of public information in Article 10 General Food Law.[247] It also appears to be inspired by the necessity to make visible the actions of the bodies responsible for achieving a higher level of food safety.[248] In particular, the Commission's decision-making process should be open to make the risk management solutions visible.[249] Such an approach should enable consumers to understand why new measures have been taken and what improvements are intended to be achieved through them. Given this context, we can conclude that there is a transparency obligation even in the crisis management provisions.

4.6.2 Application of the Transparency Dimensions

From the analysis of the provisions not touched by the New Regulation, we can notice that several provide transparency obligations. Moving to the dimensions of transparency mentioned above, we apply the framework to the obligations found in the provisions.

The *democratic dimension* of transparency appears to be present in the provisions of Articles 9 and 10 General Food Law. These principles of transparency inspire the General Food Law. Both the public consultations and information that the authorities should give to citizens aim to improve the institutions' visibility. Moreover, they enhance the *legitimacy* of the public authorities.

---

[245] Commission Implementing Decision (EU) 2019/300 of 19 February 2019 establishing a general plan for crisis management in the field of the safety of food and feed, C/2019/1064, OJ L 50, 21.2.2019, which repealed Commission Decision 2004/478/EC of 29 April 2004 concerning the adoption of a general plan for food/feed crisis management.

[246] The preliminary contact with stakeholders is particularly relevant when information is released in relation to a specific commercial brand or name. See Fuentes (2010), p. 209.

[247] Fuentes (2010), p. 209.

[248] Bolognini (2003), p. 453.

[249] Bolognini (2003), p. 453.

Another obligation to improve democracy that falls on the public authority is in the RASFF. The procedure initiated by the authorities in the network is indeed based on communication and evaluation among the different parties.

The *legitimacy* dimension may be found even in the case of RASFF and in the general plan for crisis management's communication strategy. Both measures aim to manage a food incident. Involving authorities, they should demonstrate the effectiveness of their actions.

Besides the principles of transparency and the RASFF, the other transparency obligations described in the previous paragraph mostly fall on the food business operators. One dimension particularly present among the general requirements' obligations is *accountability*. In this context, accountability qualifies the actor's performance in its responsibilities towards consumers. For instance, in Articles 14 and 16 the obligations to inform consumers on the use of the product and the ones deriving from the presentations make visible the duties of the food business operators. The accountability dimension is also evident in the obligations laid down in Article 17, whereas transparency concerns the internal organisation and the communication made by Member States, and in Article 19 on the information to consumers on the reasons for the withdrawal. Moreover, the traceability system outlined in Article 18 allows for the identification of food safety problems and the business responsible. Therefore, traceability enhances accountability.

Transparency as a means to realise *individual rights* stems from the described obligations of the General Food Law. It can be noted that the general requirements of food take into account the consumers' right to informed choices, namely the right to be informed about the health risks deriving from food.[250] This is of particular evidence in Article 14, which aims to protect the public health and the consumer from food that is either a health risk or unacceptable for consumption. The obligations of food companies to inform the authorities and the public of health risks can be found in other provisions. Besides Article 14, information to consumers carried through labelling, advertising and presentation, pursuant to Article 16, shall not mislead the consumer. Therefore, it fulfils the individual right of the consumer to receive fair information. In Article 19, the information to consumers concerns the reasons for the withdrawal of a product non-compliant with the food safety requirements. Even in this case, the obligation aims to satisfy the consumers' right to be informed.

Instead, in Article 17, the obligation to inform the consumers falls on the Member States,[251] whilst the first paragraph provides that the food business operators should ensure compliance with food law requirements.

---

[250] Fuentes (2010), p. 205.

[251] Particularly, Art. 17 (2) General Food Law. For the interpretation of this provision in connection with Article 10 General Food Law, see Judgment of the Court (Fourth Chamber), 11 April 2013, *Karl Berger v Freistaat Bayern*, C-636/11, ECLI:EU:C:2013:227, and the comment by Purnhagen (2013); Bolognini (2013), p. 164 ff. The Court of Justice of the European Union clarify the extent of Article 10 in the Karl Berger case: it shall be interpreted as "not precluding national legislation allowing information to be issued to the public mentioning the name of a food and the name or

As regards the individual rights dimensions, it has been argued that Articles 2 and 8 of the European Convention of Human Rights (ECHR), as interpreted by the ECtHR in the *Guerra* case and *Oneryildiz* case, imply the obligation of authorities to inform consumers of any serious health risk of which they may be aware, *including risks caused by food.*[252] In particular, the Court argued that in Article 8 ECHR about the respect of private and family life, an obligation may exist to inform about health risks.[253] In a similar way, the right to life disposed in Article 2 ECHR was breached in the *Oneryildiz* case because of a lack of information on environmental risks.[254] Although both cases refer to environmental risks, it appears consistent to apply them also to food safety risks. Consequently, authorities that possess information about food safety risks are under obligation to inform the consumers adequately.[255]

From the description and the analysis of the transparency obligations arising from the General Food Law, we can also conclude that all of them appear to present the *market efficiency* dimension.

For instance, the requirements of Article 14 correspond to the need to establish a general requirement for only safe food to be placed on the market so that the market functions effectively. Moreover, the rules on presentation show a link with the placing on the market of foodstuffs. The availability of information facilitates consumers' purchases and trade relations.

The same can also be said for the obligations concerning the responsibilities of food business operators in Articles 17 and 19 of the General Food Law. These obligations are intended to ensure compliance with food law and to highlight the businesses' duties within the market context. Additionally, transparency duties related to traceability consists in providing information in business relations and to authorities upon request. As such, they contribute to the development of a level playing field throughout the European market.

In this context, the RASFF system improves market transparency by creating a network for the notification of a direct or indirect risk to human health deriving from

---

trade name of the food manufacturer, processor or distributor, in a case where that food, though not injurious to health, is unfit for human consumption".

252 Van der Meulen (2007), pp. 274 ff.

253 See Council of Europe/European Court of Human Rights (2025). Van der Meulen (2007) p. 274 quotes the Court's argument: "The Court reiterates that severe environmental pollution may affect individuals' well-being and prevent them from enjoying their homes in such a way as to affect their private and family life adversely (...). In the instant case the applicants waited, right up until the production of fertilizers ceased in 1994, for essential information that would have enabled them to assess the risks they and their families might run if they continued to live at Manfredonia, a town particularly exposed to danger in the event of an accident at the factory. The Court holds, therefore, that the respondent State did not fulfil its obligation to secure the applicants' right to respect for their private and family life, in breach of Article 8 of the Convention. There has consequently been a violation of that provision".

254 In the Oneryildiz case the Court held that this reasoning also applies to the right to life protected in Article 2 ECHR. See Van der Meulen (2007), p. 274.

255 Van der Meulen (2007), p. 275; Kocharov (2009), p. 344.

food or feed. Thus, the network of authorities and the procedure itself operate in the market setting. They aim at limiting the placing on the market of unsafe food or at withdrawing and recalling it in order to protect human health, including border rejection related to health risk. Consequently, the RASFF and procedures provided in the general plan for crisis management improve market efficiency.

To conclude, the democracy and legitimacy dimensions seem to occur primarily in relation to public authorities' obligations, whereas accountability concerns both public authorities and food business operators. Although it is possible to trace all the dimensions of transparency within the examined obligations, the realisation of the individual right to informed choices and the market efficiency dimensions appear to prevail.

4.7 Analysis of the CJEU Case Law about Transparency in the Food Law Domain

4.7.1 Introduction

The overarching topic of transparency calls to observe several aspects, such as whether and how the Court of Justice of the European Union's jurisprudence has addressed transparency in the food law domain.

It should be noted that Article 19 (1) TEU defines the CJEU's role stating that "it shall ensure that in the interpretation and application of the Treaties the law is observed".[256] Accordingly, the Court has developed "principles of a constitutional nature" as part of EU law, constraining the EU institutions and Member States which act within EU law.[257] In judicial practice, the Court uses classical methods of interpretation which take into account "the wording of the provision, its context and purpose".[258] We should underline that its approach to interpretation is "purposive" or "teleological" under a dynamic perspective on the CJEU's role.[259] In this vein, the Court "examines the whole context in which a particular provision is situated" and "gives the interpretation most likely to further what the Court considers that

[256] Craig and De Búrca (2020), p. 87.

[257] As stated by Craig and De Búrca (2020), p. 93, the CJEU "fashioned seminal principles of the EU legal order, such as direct effect, supremacy, and state liability in damages" which "have defined the very nature of the EU, constitutionalizing it and distinguishing it from other international treaties".

[258] Osiejewicz (2017), p. 366. According to the Author, "[T]these classical methods of interpretation may be divided into four groups: (1) linguistic method, based on the wording of the provisions; (2) historical method, referring to the origin of the provision; (3) systemic method, taking into account the place of the provision; (4) teleological (purposive) method, which refers to the purpose of the provision" (p. 366).

[259] Craig and De Búrca (2020), p. 93.

provision sought to achieve".[260] Nonetheless, the Court's institutional point of view differs from the standpoint of national courts, EFSA, national food safety authorities, and the Commission.[261]

Another preliminary clarification concerns the fact that the Transparency Regulation has been brought to the CJEU's attention only recently. As stated before, the Transparency Regulation has been applicable only since 27 March 2021. The analysis of case law shows that the Transparency Regulation has been considered in the case *Ascenza Agro, SA and Industrias Afrasa, SA v European Commission.*[262] The applicants invoked the breach of transparency obligations within the legal framework governing the adoption of Commission Implementing Regulation (EU) 2020/17 of 10 January 2020 concerning the non-renewal of the approval of the active substance chlorpyrifos-methyl, in accordance with Regulation (EC) No 1107/2009 on the placing of plant protection products on the market ("PPPs"). The General Court stated that compliance with the transparency requirement in the field of PPPs is guaranteed by specific provisions, such as Recital 40 General Food Law and Recital 12 Regulation (EC) No 1107/2009 on the transparency of the active substance evaluation process. However, the Court's reasoning predominantly focuses on the rights related to transparency obligations, without providing substantive explanations regarding the content of these obligations and of transparency as a principle in clearly defined terms.

4.7.2 Methodology

Dealing with "multidimensional" concepts such as transparency can be challenging when it comes to judicial legal research. In the previous Chapter we clarified that the principle of openness emerged in many fields of EU law.[263] Therefore, to study the judicial perspective, Computer Assisted Qualitative Document Analysis Software (CAQDAS) was used as an effective tool.[264] Software supported content analysis enhances the systematicity of case law analysis combined with the researcher's interpretative capacities.[265] MAXQDA was used as a supported content analysis software. More in detail, CAQDAS platforms let specify the number of judgments examined based on the chosen keywords and keep track of the analysis process.

[260] Craig and De Búrca (2020), p. 94.

[261] For instance, the Commission is involved in law-making and implementing acts. On the role of the national courts exercising the judicial review of EU Law, see Osiejewicz (2017), pp. 362–365.

[262] Judgment of the General Court (Seventh Chamber, Extended Composition) of 4 October 2023. *Ascenza Agro, SA and Industrias Afrasa, SA v European Commission.* Case T-77/20. ECLI:EU:T:2023:602.

[263] See Chap. 3, para. 3.1.

[264] Schebesta (2018), pp. 23–33. For a general reflection about methodology in European legal scholarship, see van Gestel and Micklitz (2014), pp. 292–316.

[265] Schebesta (2018), p. 25.

Observations on jurisprudence are more accurate when the actual number of cases gathered is known in advance.[266] It has been argued that systematic content analysis may contribute to the empirical research process ensuring an epistemological fit between research question and methodology. Moreover, the subsequent stages of selecting and coding source documents and analysing research results strengthen the *modus operandi*.[267]

According to scholars, it is possible to find the *epistemological fit* between research question and methodology when investigating "(1) the bare outcome of legal disputes, (2) *the legal principles one can extrapolate from those outcomes*, and (3) the facts and reasons that contribute to those outcomes and principles".[268] It has been argued that "content analysis is perfectly suited for examining an aspect of judicial method", and "questions researchers have pursued include the types of authorities judges cite in their opinions; the argumentative, interpretative, or expressive techniques judges use in different circumstances".[269] Through content analysis, researchers can tackle research questions that require a descriptive methodology. To this effect, our research design encompasses *how the CJEU addresses the transparency concept in its case law related to food* (research question, RQ) by using software that supports qualitative content analysis. The question can be answered according to textual interpretation.

In *selecting the source documents*, scholars must decide their study's scope by establishing how the underlying dataset has been collected.[270] For instance, primary sources' identification includes specifying which databases and selection methods were used to spot cases.[271] CAQDAS programs help manage and organise source documents after they have been gathered and downloaded by other databases. Since MAXQDA does not support document collection, the primary documents must be imported into MAXQDA from an external database. However, source documents can be organised in different folders. We used the "Advanced Search" tool in the *Eur-lex* website[272] as an external database. By means of search terms within the CJEU case law and by narrowing the results to "judgments", we used the keywords

[266] According to Schebesta (2018), p. 24, "descriptive observations about jurisprudence should increase in accuracy with the number of observations approaching the actual number of cases", therefore "quantity becomes quality" of the research.

[267] Schebesta (2018), p. 25. On empirical legal research, see also the references at p. 25, note 16, particularly Cane and Kritzer (2010).

[268] Hall and Wright (2008), p. 85 as also referred by Schebesta (2018), p. 26 (emphasis added).

[269] Hall and Wright (2008), p. 93; Schebesta (2018), p. 26.

[270] Schebesta (2018), p. 26.

[271] As observed by Schebesta (2018), p. 26, "[…] in EU Law, on often finds studies in which it is unclear which cases, or how many, were identified and through which parameters. For the sake of transparency, accountability and replicability, legal researcher must systematically start to account for the identification and selection of primary sources used in given legal research".

[272] https://eur-lex.europa.eu/advanced-search-form.html.

"transparen*" AND "food".[273] An initial analysis conducted up to January 2022 identified 185 case law, which were imported into MAXQDA. The case law analysis presented in this chapter began by collecting decisions issued up to January 2022, providing a comprehensive review of the relevant judicial precedents from that period. Additionally, a total of 38 new cases were reviewed as part of an updated analysis, incorporating case law developments from January 2022 to December 2024. Once the documents are collected, *coding source documents means* assigning a code to the data as a "classification device" using a category, a description, or a direct quote from the source.[274] The coding process aims to create a set of codes based on the given research question (RQ), arguably forcing the researcher to be more precise about their decisions in interpreting data.[275] All the documents were reviewed separately and systematically in order to seize and later analyse the case law. After discovering the sentences related to "transparen*" and "food" in the selected document's text using the searching tool,[276] we attributed *bottom-up coding*.[277] Therefore, the coding process was not predefined, and the chosen *descriptive coding* was derived inductively from the judgments.

After an initial exploratory coding phase, we discovered that only 63 of 223 case law concern the food law domain. The remaining cases were related to various issues such as state aid, transparency in tender procedures, transparency in the market or competition domain, antidumping, import, and access to documents in general not associated with food law. These cases were grouped in a MAXQDA document folder named "disregarded" to separate them from the relevant ones. Therefore, all these judgments fell out of the scope of the research and have not been coded.

From the explanatory coding phase of the 63 case law intertwined with food law, none of them seemed to unveil a CJEU's reasoning on an overarching "principle of transparency". Nonetheless, we have coded the data with categories and descriptions to verify this assumption and, more generally, the judicial acknowledgement of transparency.

The *coding case outcome* was carried out using the following codes: legal provisions; connection with the principle of legal certainty; connection with the principle of sound administration; traceability; policy in the BSE crisis; information to

[273] The advanced search tool in Eur-lex allows to insert the Boolean operators "AND"/ "OR" and "NOT".

[274] See, also for the different coding techniques of particular relevance in legal analysis (such as "attribute coding", "magnitude coding", "simultaneous coding", "structural coding" and "descriptive or topic coding"), see Schebesta (2018), p. 27.

[275] Usually "two cycles of coding are regarded as necessary: an initial exploratory coding phase and a systematic one" (Schebesta (2018), p. 27).

[276] All four main windows in MAXQDA have a magnifying glass icon in their toolbars.

[277] The *top-down* coding includes predefined codes based on a theory to scientifically test and falsify hypothesis. On the contrary, the *bottom-up coding* has been attached to grounded theory which is used "in order to infer theory during the coding process and in interaction with the data". See Schebesta (2018), p. 29.

consumer on the characteristics of goods; general requirements of transparency; transparency and right to protection of personal data; risk management; risk assessment principle; inconsequential use (see Fig. 4.1).

The *analysis of the coding process results* aims to "explore theoretical concepts and provide support for theoretical developments by using a research tool that shows *patterns or relationships within the data*".[278] Although qualitative content analysis does not offer a different legal data analysis approach, some recommend it for theory-guided text analysis.[279] It has been noted that all lawyers go "through a process of implied *mental coding*"[280] when analysing legal documents, for example, by observing the evolution of courts' interpretations over time. However, software supported content analysis aids in the systematicity and methodological consistency of the research.[281] Therefore, uncovering connections within the data has been valuable for this purpose.

4.7.3 Analysing Cases

It is possible to find segments related to a specific topic by assigning codes to appropriate document pieces. Using the "retrieving documents" function in MAXQDA, segments previously labelled under the same code can be called up.[282] This approach allows for easy visualisation and connection of relevant segments. To identify the Court's approach to transparency, the cases will be described according to the categories.

4.7.3.1 Transparency as Mentioned in Legal Provisions' Text That the CJEU Merely Recalls in Its Judgments

This category pertains to cases where transparency is merely mentioned in the wording of the legal provision. While coding the documents, we decided to assign as a code the name of the article to the segment related to a specific legal provision the Court has recalled in its judgments. Subsequently, we collected all the legal

---

[278] Schebesta (2018), p. 28. Emphasis added. According to the author, "one of the primary goals of coding is to identify patterns (or the lack thereof) in the data, characterized by, for instance, similarity, differences, frequency, sequence, correspondence or causation".

[279] See Mayring (2014) recalled by Schebesta (2018), p. 28.

[280] Schebesta (2018), p. 28. Emphasis added.

[281] Schebesta (2018), p. 28; Hall and Wright (2008), p. 64.

[282] In MAXQDA, the retrieval of the segment is based on a function called "activation" of documents and codes. Namely, all documents' segments from "activated documents" that were coded with any of the "activated codes" are called up and shown in the "Retrieved segments" windows (see the Online Manual – MAXQDA 2024, "The Principle of Segment Retrieval").

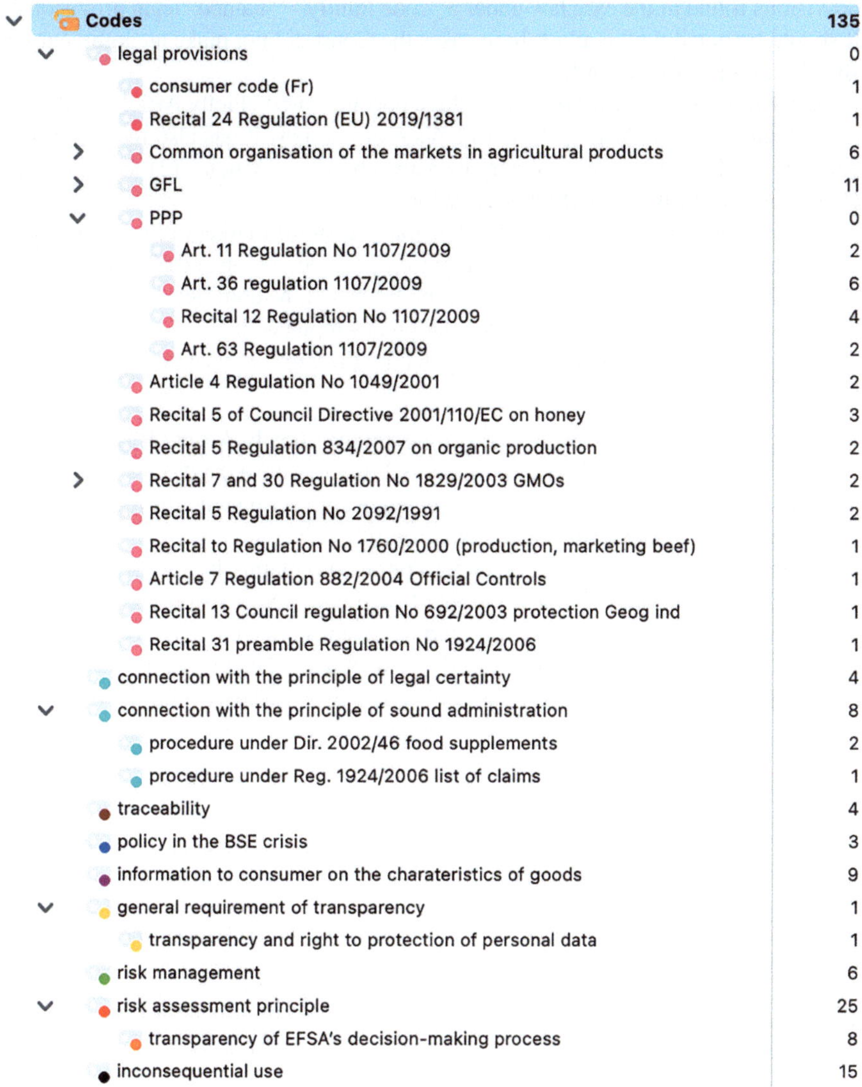

Fig. 4.1 Hierarchical code system structure (MAXQDA 2024)

provisions found in the case law under a "code family"[283] named "legal provisions". It should be noted that in all of these cases the Court did not make an argument but simply cited the articles of law.

Regarding the General Food Law, the Court has cited chiefly Article 6 GFL on the risk analysis and Article 38 on EFSA's transparency. It has also recalled transparency as a feature of precautionary measures based on Article 7(2). However, the Court does not further specify this assertion.

The Regulation (EC) No 1107/2009 on placing on the market of Plant Protection Products ("PPPs")[284] has been quoted regarding Articles 11, 36 and 63. Interestingly, the latter concerns the confidentiality of the information submitted according to the Regulation, and the Transparency Regulation has replaced it.

Other legal provisions and recitals quoted by the Court in its case law regard, for instance:

- transparency referred to access to documents (specifically, Article 4(2) Regulation (EC) No 1049/2001 as a base for the possibility to refuse the disclosure of documents relating to toxicity studies conducted in connection with the renewal of the approval of the active substance glyphosate)[285];
- "a transparent […] framework underpinning the claims by the products concerned to be of quality"[286] (Recital 92 of Regulation No 1308/2013[287]);
- full transparency concerning "the particular consumer interests as regards the geographical characteristics of honey"[288] (Recital 5 of Directive 2001/110/EC[289]);
- the "interest of consumer to receive adequate and transparent product information" in Article 113(2)(a)(iii) of the Single CMO Regulation[290];

[283] Through the use of the software, it is possible to collect codes and create sub-codes, organizing the way of coding in order to have a clearer framework.

[284] Regulation (EC) No 1107/2009 of the European Parliament and of the Council of 21 October 2009 concerning the placing of plant protection products on the market and repealing Council Directives 79/117/EEC and 91/414/EEC, OJ L 309, 24.11.2009, p. 1–50.

[285] For instance, Judgment of the General Court (Eighth Chamber) of 7 March 2019, *Antony C. Tweedale v European Food Safety Authority*, Case T-716/14, ECLI:EU:T:2019:141.

[286] Judgment of the Court (Second Chamber) of 20 December 2017, *Comité Interprofessionnel du Vin de Champagne v Aldi Süd Dienstleistungs-GmbH & Co.OHG*, C-393/16, ECLI:EU:C:2017:991.

[287] Regulation (EU) No 1308/2013 of the European Parliament and of the Council of 17 December 2013 establishing a common organisation of the markets in agricultural products and repealing Council Regulations (EEC) No 922/72, (EEC) No 234/79, (EC) No 1037/2001 and (EC) No 1234/2007, OJ L 347, 20.12.2013.

[288] Judgment of the Court (Third Chamber) of 22 September 2016, *Breitsamer und Ulrich GmbH & Co. KG v Landeshauptstadt München*, C-113/15, ECLI:EU:C:2016:718.

[289] Council Directive 2001/110/EC of 20 December 2001 relating to honey, OJ L 10, 12.1.2002.

[290] Judgment of the General Court (Eighth Chamber), 13 November 2014, *Kingdom of Spain v European Commission*, T-481/11, ECLI:EU:T:2014:945. Regulation (EU) No 1308/2013 of the European Parliament and of the Council of 17 December 2013 establishing a common organisation of the markets in agricultural products and repealing Council Regulations (EEC) No 922/72, (EEC) No 234/79, (EC) No 1037/2001 and (EC) No 1234/2007, OJ L 347, 20.12.2013.

- Recital 5 of Regulation (EC) No 834/2007 on organic production, which provides that it "is [...] appropriate to define more explicitly the objectives, principles and rules applicable to organic production, to contribute to transparency and consumer confidence"[291];
- the necessity of a "transparent [...] authorisation procedure" before placing on the market food and feed consisting of, containing or produced from genetically modified organisms ("GMOs") in the preamble to Regulation No 1829/2003[292];
- the improvement of the "transparency of the conditions for the production and marketing" of the beef products (Recitals (4) to (7) in the preamble to Regulation No 1760/2000[293]);
- Article 7 Regulation (EC) No 882/2004 on official controls, which stated that "the competent authorities shall ensure that they carry out their activities with a high level of transparency"[294];

---

[291] Council Regulation (EC) No 834/2007 of 28 June 2007 on organic production and labelling of organic products and repealing Regulation (EEC) No 2092/91, OJ L 189, 20.7.2007. The Regulation is now repealed by Regulation (EU) 2018/848 of the European Parliament and of the Council of 30 May 2018 on organic production and labelling of organic products and repealing Council Regulation (EC) No 834/2007, PE/62/2017/REV/1, OJ L 150, 14.6.2018. On the same topic, two case law recalled the fifth Recital of the Council Regulation (EEC) No 2092/91 of 24 June 1991 on organic production of agricultural products and indications referring thereto on agricultural products and foodstuffs, OJ L 198, 22.7.1991, repealed by Council Regulation (EC) No 834/2007, which stated that "it is therefore appropriate to define more explicitly the objectives, principles and rules applicable to organic production, in order to contribute to transparency and consumer confidence as well as to a harmonised perception of the concept of organic production". See Judgment of the Court (Ninth Chamber) of 12 October 2017, *Kamin und Grill Shop GmbH v Zentrale zur Bekämpfung unlauteren Wettbewerbs Frankfurt am Main eV*, C-289/16, ECLI:EU:C:2017:758.

[292] Regulation (EC) No 1829/2003 of the European Parliament and of the Council of 22 September 2003 on genetically modified food and feed (Text with EEA relevance), OJ L 268, 18.10.2003. In the same vein, one case law refers to Article L. 531-2-1of the French Code de l'environnement (Environmental Code). See, for instance, Judgment of the Court (Grand Chamber) of 25 July 2018, *Confédération paysanne and Others v Premier ministre and Ministre de l'agriculture, de l'agroalimentaire et de la forêt*, C-528/16, ECLI:EU:C:2018:583.

[293] Judgment of the Court (Second Chamber) of 24 May 2007. *Maatschap Schonewille-Prins v Minister van Landbouw, Natuur en Voedselkwaliteit,* C-45/05, ECLI:EU:C:2007:296 refers to Regulation (EC) No 1760/2000 of the European Parliament and of the Council of 17 July 2000 establishing a system for the identification and registration of bovine animals and regarding the labelling of beef and beef products and repealing Council Regulation (EC) No 820/97, OJ L 204, 11.8.2000.

[294] Judgment of the Court (Fourth Chamber), 11 April 2013, *Karl Berger v Freistaat Bayern*, C-636/11, ECLI:EU:C:2013:227 concerning Regulation (EC) No 882/2004 of the European Parliament and of the Council of 29 April 2004 on official controls performed to ensure the verification of compliance with feed and food law, animal health and animal welfare rules, OJ L 165, 30.4.2004 replaced by Regulation (EU) 2017/625 of the European Parliament and of the Council of 15 March 2017 on official controls and other official activities performed to ensure the application of food and feed law, rules on animal health and welfare, plant health and plant protection products, amending Regulations (EC) No 999/2001, (EC) No 396/2005, (EC) No 1069/2009, (EC) No 1107/2009, (EU) No 1151/2012, (EU) No 652/2014, (EU) 2016/429 and (EU) 2016/2031 of the European Parliament and of the Council, Council Regulations (EC) No 1/2005 and (EC) No

- generic "transparency reasons" why "the simplified procedure provided for in Article 17 of Council Regulation (EEC) No 2081/92 for the registration of names already protected or established by usage in the Member States does not provide for any right of objection" shoud be deleted[295];
- Recital 31 in the preamble to Regulation (EC) No 1924/2006, stating that the Register of permitted health claims is a means of communication for the purposes of transparency and information in order to avoid requests concerning claims.[296]

In all these cases, the Court merely quotes the articles or the recitals. Therefore, from these segments it is not possible to draw any meaningful conclusion on the transparency principle.

4.7.3.2 Transparency Is Connected with Other Legal Principles

Chapter 3 examined the principle of transparency in EU law and observed that it has developed alongside other well-established legal principles, rather than standing as an overarching principle. The CJEU in the *Natural Health* case has indeed referred to the connection between transparency and sound administration in a case about the legitimacy of the food supplements' authorisation procedure.[297] Moreover, in the *Consorzio del Prosciutto di Parma* case, the Court has argued against a possible breach of both legal certainty and transparency principle.[298] It clarified in particular that the scope and effect of a European law provision must be clear and foreseeable to individuals. Otherwise, both the principle of legal certainty and the principle of transparency would have been breached.

1099/2009 and Council Directives 98/58/EC, 1999/74/EC, 2007/43/EC, 2008/119/EC and 2008/120/EC, and repealing Regulations (EC) No 854/2004 and (EC) No 882/2004 of the European Parliament and of the Council, Council Directives 89/608/EEC, 89/662/EEC, 90/425/EEC, 91/496/EEC, 96/23/EC, 96/93/EC and 97/78/EC and Council Decision 92/438/EEC (Official Controls Regulation), OJ L 95, 7.4.2017. Article 7 Regulation (EC) No 882/2004 has been replaced by Article 11 Regulation (EU) 2017/625 on "Transparency of official controls".

[295] Judgment of the Court (Fourth Chamber) of 2 July 2009, *Bavaria NV and Bavaria Italia Srl v Bayerischer Brauerbund eV*, C-343/07, ECLI:EU:C:2009:415.

[296] Judgment of the General Court (Eighth Chamber) of 12 June 2015, *Plantavis GmbH and NEM, Verband mittelständischer europäischer Hersteller und Distributoren von Nahrungsergänzungsmitteln & Gesundheitsprodukten eV v European Commission and European Food Safety Authority*, T-334/12, ECLI:EU:T:2015:376.

[297] Judgment of the Court (Grand Chamber) of 12 July 2005, *The Queen, on the application of Alliance for Natural Health and Nutri-Link Ltd v Secretary of State for Health* (C-154/04) and *The Queen, on the application of National Association of Health Stores and Health Food Manufacturers Ltd v Secretary of State for Health and National Assembly for Wales* (C-155/04), Joined cases C-154/04 and C-155/04, ECLI:EU:C:2005:449. See Chap. 3, para. 3.2.1.

[298] Judgment of the Court of 20 May 2003, *Consorzio del Prosciutto di Parma and Salumificio S. Rita SpA v Asda Stores Ltd and Hygrade Foods Ltd.*, Case C-108/01, ECLI:EU:C:2003:296. See Chap. 3, para. 3.2.2.

Interestingly, both cases just mentioned also resulted from our analysis through CAQDAS. We assigned codes highlighting the connection of transparency with both legal certainty and sound administration. As seen in *Natural Health,* the procedure for authorising food supplements under Directive 2002/46/EC was deemed to be lacking in transparency. This was allegedly due to a "lack of precision" in the criteria applied by the European Food Safety Authority.[299] The Court acknowledged that the absence of those provisions should be compensated by the Commission that had to adopt the measure "in accordance with the *principle of sound administration*" in order to ensure that the consultation stage with the European Food Safety Authority is carried out "*transparently* and within a reasonable time".[300] However, the Court's deductions on these statements are limited, and they do not further express any reasoning on transparency. This can be seen from the Court's "principle-consistent interpretation"[301] of the Directive 2002/46/EC, which points out that the Commission's executive powers must be seen in the light of the sound administration and legal certainty principles. A "transparent procedure" is deemed to be able to carry out these legal principles.

In a similar way, in the case *Health Food Manufacturers* the Court rejected the applicant's line of argument that alleged the infringement of the principle of sound administration due to the "lack of transparency" on the part of the Commission.[302] The plea specifically concerned the Commission's decision to adopt a list of permitted claims under the procedure laid down by Article 13(3) Regulation (EC) No 1924/2006 on Nutrition and Health Claims ("NHCR").[303] Even if the CJEU's rejection was based on the facts of the case, the Court stressed that the Commission's decision was taken "in a transparent manner" by setting out "in a clear manner the reasoning behind its decision to adopt the list of permitted claims".[304] It follows that

[299] Judgment of the Court (Grand Chamber) of 12 July 2005, Joined cases C-154/04 and C-155/04, para. 71–72: "The claimants in the main action submit that the procedures referred to in Article 4 (5) and (6) of Directive 2002/46 lack transparency because of the lack of precision in the criteria applied by the European Food Safety Authority in its examination of dossiers seeking authorisation to use a substance not included on the positive lists. The procedures thus represent a particularly heavy financial and administrative burden. In that regard, a measure which, like that at issue in the main actions, includes a prohibition on marketing products containing substances not included on the positive lists laid down in the applicable legislation must be accompanied by a procedure designed to allow a given substance to be added to those lists and the procedure must comply with the general principles of Community law, in particular the principle of sound administration and legal certainty".

[300] Judgment of the Court (Grand Chamber) of 12 July 2005, Joined cases C-154/04 and C-155/04, para. 81.

[301] See Vandamme (2006), p. 314. See Chap. 3, para. 3.2.1.

[302] Judgment of the General Court (Eighth Chamber) of 12 June 2015, *The Health Food Manufacturers'Association and Others v European Commission*, T-296/12, ECLI:EU:T:2015:375.

[303] Regulation (EC) No 1924/2006 of the European Parliament and of the Council of 20 December 2006 on nutrition and health claims made on foods, OJ 404, 30.12.2006, pp. 9–25.

[304] Judgment of the General Court (Eighth Chamber) of 12 June 2015, *The Health Food Manufacturers' Association and Others v European Commission*, para. 103 and 109. For some considerations on the NHCR, see Merten-Lentz (2019), pp. 13–24; La Manna (2013), pp. 313 ff.

there was no breach of the sound administration principle. As to the emergence of transparency next to established legal principles, it seems possible to confirm what was previously observed about transparency from a broader EU law perspective.

4.7.3.3 Transparency as a Traceability's Feature

In *Commission v. France* (C-1/00), we found a "transparent traceability" concept coded in the system as "traceability".[305] At a closer look, the named traceability concerns meat products dispatched from the United Kingdom to France under the Date-based Export Scheme ("DBES") set out in Council Decision 98/256. In this context, transparency seems to be a mere attribute of the traceability system. Namely, it is interpreted as a fundamental condition for the operations of the scheme set to protect public health. The particularity of the case does not allow to reach any conclusion on transparency as a traceability feature in the CJEU's view.

4.7.3.4 Lack of Transparency in the BSE Crisis

In some rulings,[306] the reference to transparency is related to the BSE crisis.[307] More in detail, the Court reports the applicants' arguments against the alleged Council and Commission's policies of opaqueness and lack of transparency.[308]

---

[305] Judgment of the Court of 13 December 2001, *Commission of the European Communities v French Republic*, C-1/00, ECLI:EU:C:2001:687.

[306] Judgment of the Court of First Instance (First Chamber) of 13 December 2006, *Julia Abad Pérez and Others v Council of the European Union and Commission of the European Communities*, T-304/01, ECLI:EU:T:2006:389; Judgment of the Court of First Instance (First Chamber) of 13 December 2006; *É. R., O. O., J. R., A. R., B. P. R. and Others v Council of the European Union and Commission of the European Communities*, T-138/03, ECLI:EU:T:2006:390; Judgment of the Court (Grand Chamber) of 6 December 2005, *The Queen, on the application of ABNA Ltd and Others v Secretary of State for Health and Food Standards Agency* (C-453/03), *Fratelli Martini & C. SpA and Cargill Srl v Ministero delle Politiche Agricole e Forestali and Others* (C-11/04), *Ferrari Mangimi Srl and Associazione nazionale tra i produttori di alimenti zootecnici (Assalzoo) v Ministero delle Politiche Agricole e Forestali and Others* (C-12/04) and *Nederlandse Vereniging Diervoederindustrie (Nevedi) v Productschap Diervoeder* (C-194/04), Joined cases C-453/03, C-11/04, C-12/04 and C-194/04, ECLI:EU:C:2005:741.

[307] On BSE crisis, see more in detail, Chap. 2, para. 2.2.1. and this chapter, para. 4.1.

[308] See, for instance, Judgment of the Court of First Instance (First Chamber) of 13 December 2006, *Julia Abad Pérez and Others v Council of the European Union and Commission of the European Communities*, para. 61: "The applicants maintain that the Commission and the Council acted in breach of the Community rules governing animal health and public health. In support of that position, they refer to the report of the Committee of Inquiry, in which the Parliament found that the Council and the Commission were liable in the BSE crisis. With respect to the Commission in particular, that report contains the following criticisms: the suspension of veterinary inspection missions in the United Kingdom between 1990 and 1994; the lack of coordination between the different competent Directorates General; *the lack of transparency in the workings of the SVC, as its BSE sub-group has almost always been presided over by a British national and made up of*

Accordingly, these Institutions were accused of infringing the legitimate expectations of European consumers in order to prevent the broadcasting of the BSE effects from causing "a collapse in the market of bovine meat".[309] The objective of transparency was also invoked by the Parliament, when it referred to the loss of credibility of the competent authorities as a result of the BSE crisis in case *ABNA Ltd and Others v Secretary of State for Health and Food Standards Agency.*[310] According to the Parliament, the obligation to indicate the percentages of feedingstuff's ingredients is a measure not sufficient to contribute to the objective of safeguarding animal and human health. As stated by the CJEU, this information made it possible to identify the ingredients of feedingstuff suspected of contamination without waiting for the results of laboratory analyses and to rapidly withdraw the feedingstuff from use.[311] Despite these judgments refer to a policy of transparency, the Court has not argued on this specific point, and it mostly ruled on the proportionality of the measures set to protect animal and human health safety.

4.7.3.5 "Transparent Information" to Consumers

It is noteworthy to look at the relationship between transparency and information to consumers. As observed, from a conceptual point of view transparency consists of the *availability and exchangeability of information* between different actors in a specific context.[312] Despite *information* is a key concept that allows us to look at the interactions between different subjects of the food chain, there is little attention to it in the case law we analysed.

Interestingly, the case *Vinegar*[313] mentions transparency as linked with information to consumers. This is one of the most well-known judgments on the internal market rules, and it can be classified in the "market-oriented phase" of food law development.[314] It concerns the Commission's infringement action against the Italian legislation prohibiting marketing and importation of products containing acetic acid not coming from the acetic fermentation of wine and limiting the name

numerous British scientists; the failure to guarantee proper execution of the veterinary checks and to comply with the obligations provided for by Directive 89/662, and the tardiness and ineffectiveness of the rules adopted on meat-and-bone meal. The applicants state that the action is brought against the Council and the Commission because they shared the competences to be exercised to deal with the BSE crisis" (emphasis added).

[309] See Judgment of the Court of First Instance (First Chamber) of 13 December 2006; *É. R., O. O., J. R., A. R., B. P. R. and Others v Council of the European Union and Commission of the European Communities*, para. 80.

[310] See Joined cases C-453/03, C-11/04, C-12/04 and C-194/04, para. 75.

[311] See Joined cases C-453/03, C-11/04, C-12/04 and C-194/04, para. 76.

[312] See Chap. 2, para. 2.3. For an analysis of the information to consumer and labelling as a "*première composante du principe de transparence*", see Bossis (2005).

[313] Judgment of the Court of 9 December 1981, *Commission of the European Communities v Italian Republic*, C-193/80, ECLI:EU:C:1981:298.

[314] See in this Chapter, para. 4.1.

"vinegar" only to products obtained from the acetic fermentation of wine. This regulation was deemed to be applicable to products of foreign origin and not to national production. According to the Court, wine-vinegar represented a typical Italian production, whereas other natural vinegars made of other fruits were mainly products of foreign origin. In this perspective, it was a clear case of *de facto* discriminatory measure.[315] As justification of its rules on this matter, the Italian Government claimed that it was necessary to protect consumers from the risk of being misled as to "the essential nature of the raw material used and of the end-product".[316] However, the CJEU argued that such protection may be provided by other means enabling national and imported products to be treated alike, for instance by labelling them with additional specific information on the type of marketed vinegar. This approach "would enable the consumer to make his choice in full knowledge of the facts and *would guarantee transparency in trading and in offers to the public* by providing an indication of the raw material used to make the vinegar".[317] Transparency, therefore, seems to relate to information to consumers in order to raise awareness on the product's characteristics.

On the back of the *Vinegar* judgment, the CJEU in case *Commission v Germany* has further stressed that the possibility to enable consumers to make informed choices may be ensured "by means which do not prevent the importation of products which have been lawfully manufactured and marketed in other Member States and, in particular, by the compulsory affixing of suitable labels giving the nature of the product sold".[318] According to the Court, indicating raw materials utilized in the manufacture of beer would guarantee transparency "in trading and in offers to the public".[319] The Court has hence expressed that the absolute ban on the marketing of beer containing additives cannot be justified on public health grounds and it constitutes a barrier to trade.[320]

---

[315] Accordingly, by prohibiting the marketing and importation of vinegars of agricultural origin other than those originating in the acetic fermentation of wine and by restricting the designation "vinegar" to wine-vinegar, Italy breached its obligation under Article 30 EEC Treaty. On *Vinegar,* see Bestagno (2020), p. 52 ff.

[316] Judgment of the Court of 9 December 1981, *Commission of the European Communities v Italian Republic*, para. 25.

[317] Judgment of the Court of 9 December 1981, *Commission of the European Communities v Italian Republic*, para. 27. According to Clark (1988) p. 767, the *Vinegar* case appears to make "the consumer protection justification under the rule of reason a dead letter in this area due to the availability of labelling requirements".

[318] Judgment of the Court of 12 March 1987, *Commission of the European Communities v Federal Republic of Germany,* C-178/84, ECLI:EU:C:1987:126, para. 35.

[319] Judgment of the Court of 12 March 1987, *Commission of the European Communities v Federal Republic of Germany,* para. 35.

[320] As regard the consequences on the liability profiles, Steiner (2007), p. 76, notes that the illegality of the rule prohibiting the use of additives, which caused the plaintiff's loss, was not sufficient clear until *Commission v Germany*: the "Germany breach was not 'sufficiently serious' to attract liability. Although it was clear before that case that the prohibition on the designation of beer for products not satisfying the beer purity requirements was in breach of Community law, there was

The CJEU has furthermore emphasised the need to ensure that the consumer has "the most accurate and *transparent information possible concerning the characteristics of good*" in the case *Neptune Distribution.*[321] Accordingly, enabling consumers to receive "transparent information" is closely related to the protection of human health and it is a question of general interest.[322] More in detail, the case *Neptune Distribution* concerns the breach of the provisions of the Nutrition and Health Claims Regulation and Directive 2009/54/EC, which lay down limitations to the use of the claims and indications at issue. Precisely, the claims related on the sodium content of natural mineral water. Although the Court failed to provide an interpretation of "transparent information", it links them with the aim of ensuring a high level of consumer protection and fair trading. The CJEU has observed that packaging or labels containing an indication that suggests mineral waters are low in sodium or salt may mislead the consumer when the total sodium content exceeds the limit provided for by EU legislation, even if claims referring to the sodium content of mineral waters associated with chloride ions "can be regarded as being substantively correct".[323] For this reason, the need to ensure "transparent information" is deemed sufficient to justify a limitation on the freedom of expression and information or the freedom to conduct a business under Articles 11 and 16 of the Charter of Fundamental Rights.

In a similar way, in case *Dextro Energy v Commission* the Court considered that the health claims at issue were incomplete and misleading because they highlighted only the beneficial effects of glucose for energy-yielding metabolism, even if this information was substantively correct.[324] The applicant argued that a "label ensuring the *transparency of offers* to consumer" might be sufficient to guarantee consumer protection.[325] However, the Court did not reply on the substance since the cited case

no direct causal link between that breach and the plaintiff's loss". For a detailed comment on the internal market profiles, see Clark (1988), p. 753 ff.

[321] Judgment of the Court (Fourth Chamber) of 17 December 2015, *Société Neptune Distribution v Ministre de l'Économie et des Finances*, C-157/14, ECLI:EU:C:2015:823, para. 74.

[322] Judgment of the Court (Fourth Chamber) of 17 December 2015, *Société Neptune Distribution v Ministre de l'Économie et des Finances*, para. 74. The Court recalls the ECtHR Judgements *Hertel v. Switzerland*, 25 August 1988, Report of Judgements and Decisions 1998-VI, § 47, and *Bergens Tidende and Others v. Norway*, no. 26132/95, ECHR 200-IV, § 51 where it has been argued that "the need to ensure the consumer has the most accurate and transparent information possible concerning the characteristics of goods is closely related to the protection of human health and is a question of general interest which may justify limitations on the freedom of expressions and information of a person carrying on a business or his freedom to conduct a business".

[323] Judgment of the Court (Fourth Chamber) of 17 December 2015, *Société Neptune Distribution v Ministre de l'Économie et des Finances*, para. 77.

[324] Judgment of the General Court (Fifth Chamber) of 16 March 2016, *Dextro Energy GmbH & Co. KG v European Commission*, T-100/15, ECLI:EU:T:2016:150, para. 78.

[325] Judgment of the General Court (Fifth Chamber) of 16 March 2016, *Dextro Energy GmbH & Co. KG v European Commission*, para. 89.

law concerned national non-harmonised measures while the case at stake was based on Regulation (EC) No 1924/2006.[326]

4.7.3.6 Transparency as a Risk Management's Feature

In the judgments analysed, transparency also appears as a risk management's feature.[327] As previously seen in the General Food Law's analysis, *risk management* concerns the adoption of policy alternatives in the light of the risk assessment's results, taking the necessary measures to prevent, reduce or eliminate risks. It can also refer to actions taken by institutions to reduce risks to levels deemed acceptable for society.

The cases highlight that the provisional measures adopted in the risk management actions' context must be *transparent.*[328] In this vein, transparency is a requirement of risk management measures. Along with it, the rulings refer to other concepts such as proportionality, non-discrimination and consistency with similar measures already taken.

4.7.3.7 Transparency as a Risk Assessment Principle

The case law considered transparency mostly with regard to risk assessment. As previously observed in the General Food Law's investigation, risk assessment is a component of the risk analysis principle and aims to provide scientific advice to risk managers. In the well-known case *Pfizer Animal Health*, the Court of First Instance stated that "a scientific risk assessment carried out as thoroughly as possible on the basis of scientific advice founded on the principles of excellence, *transparency* and independence is an important *procedural guarantee* whose purpose is to ensure the scientific objectivity of the measures adopted and preclude any arbitrary

[326] Judgment of the General Court (Fifth Chamber) of 16 March 2016, *Dextro Energy GmbH & Co. KG v European Commission,* para. 90.

[327] Judgment of the General Court (First Chamber, Extended Composition) of 17 May 2018, *BASF Agro BV and Others v European Commission*, T-584/13, ECLI:EU:T:2018:279; Judgment of the General Court (Eighth Chamber) of 12 June 2015, *The Health Food Manufacturers' Association and Others v European Commission*, Judgment of the Court (Third Chamber) of 1 April 2004, *Bellio F.lli Srl v Prefettura di Treviso*, C-286/02, ECLI:EU:C:2004:212. See also Judgment of the General Court (Sixth Chamber, Extended Composition) of 14 September 2022, *Pollinis France v European Commission.* Joined Cases T-371/20 and T-554/20, ECLI:EU:T:2022:556 and Judgment of the General Court (Fourth Chamber) of 21 February 2024, *Pesticide Action Network Europe (PAN Europe) v European Commission*, T-536/22, ECLI:EU:T:2024:98.

[328] See, more in detail, Judgment of the Court (Third Chamber) of 1 April 2004, *Bellio F.lli Srl v Prefettura di Treviso*, para. 58–59 and Judgment of the General Court (First Chamber, Extended Composition) of 17 May 2018, *BASF Agro BV and Others v European Commission*, para 74–75.

measures".[329] This statement has been consistently repeated in subsequent rulings in the field of food law as a result from our analysis. *Pfizer Animal Health* deals with the precautionary measure of not allowing the use of a certain type of antibiotic in animal feed in order to protect human health. The CFI argued on the principles of risk assessment and scientific committees' discretion in deciding which protective measures to adopt. Specifically, the Court tackled the precautionary principle for the first time and confirmed what the Commission had laid down in its Communication on the Precautionary Principle of 2000.[330] However, the CFI's approach has been criticised for not applying its own constraints in its ruling.[331] This remark addresses in particular the scientific advice's fulfilment of the *excellence, independence and transparency* principles that the CFI views as of "the utmost importance in risk assessment".[332] Nevertheless, the Court has not further interpreted the transparency principle but highlights that it entails a *procedural guarantee*.

The statement expressed by the CFI in *Pfizer* has been echoed in judgments regarding the procedures about plant protection products under Regulation (EC) No 1107/2009.[333] In case *ClientEarth and Pesticide Action Network Europe (PAN Europe) v European Food Safety Authority*,[334] the CJEU further specifies that "the *transparency of the process*" followed by a public authority for the adoption of a measure "likely to have an impact on the activities of economic operators" contributes to the authority "acquiring *greater legitimacy* in the eyes of the persons to whom that measure is addressed and increasing their confidence in that authority, and to ensuring that the authority is more *accountable to citizens* in a *democratic* system".[335] In this context, the CJEU has relied on the need to ensure a *transparent*

[329] Judgment of the Court of First Instance (Third Chamber) of 11 September 2002, *Pfizer Animal Health SA v Council of the European Union*, T-13/99, ECLI:EU:T:2002:209, para. 172.

[330] See Vos (2004), pp. 187 ff.; Pugliese (2017), pp. 8 ff. On the precautionary principle, see para. 4.2.2.2. According to Ellen Vos, even if the CFI followed the Communication on precautionary principle's criteria, it subsequently failed to verify the risk assessment accordingly (p. 192). A similar key case about antibiotics in feedstuffs and precautionary principle is Judgment of the Court of First Instance (Third Chamber) of 11 September 2002, *Alpharma Inc. v Council of the European Union*, T-70/99, ECLI:EU:T:2002:210.

[331] Vos (2004), pp. 192–199. According to the Author, "it seems a missed opportunity for the CFI to set high standards as regards the quality of the alternative scientific evidence relied on, in terms of excellence, independence and transparency" (p. 199). See also Douma (2003), pp. 372–405.

[332] Judgment of the Court of First Instance (Third Chamber) of 11 September 2002, *Pfizer Animal Health SA v Council of the European Union*, para. 268.

[333] For instance, Judgment of the General Court (Fourth Chamber) of 14 December 2018, *Arysta LifeScience Netherlands BV, formerly Chemtura Netherlands BV v European Food Safety Authority*, T-725/15, ECLI:EU:T:2018:977; Judgment of the General Court (First Chamber, Extended Composition) of 17 May 2018, *Bayer CropScience AG and Others v European Commission*, Cases T-429/13 and T-451/13, ECLI:EU:T:2018:280.

[334] Judgment of the Court (Second Chamber) of 16 July 2015, *ClientEarth and Pesticide Action Network Europe (PAN Europe) v European Food Safety Authority*, Case C-615/13, ECLI:EU:C:2015:489.

[335] Judgment of the Court (Second Chamber) of 16 July 2015, *ClientEarth and Pesticide Action Network Europe (PAN Europe) v European Food Safety Authority*, para. 56. Emphasis added. See

decision-making process of the European Food Safety Authority. Nevertheless, the Court has emphasised the role of institutional responsiveness, arguing that transparency is a guarantee for accountability, legitimacy and democracy.

4.7.3.8 Transparency and Personal Data

Besides the profile focused on risk assessment mentioned in the previous paragraph, in *ClientEarth and PAN Europe v EFSA* the Court also addressed transparency in the personal data field under Regulation (EC) No 1049/2001 and Regulation (EC) No 45/2001.[336] The case concerned the appeal of two NGOs against the judgment of the General Court. The General Court dismissed their action for the annulment of EFSA's refusal to disclose the names of the external experts who had collaborated to the guidance document for operators wanting to submit an application for authorisation to place plant protection products on the market. In particular, the applicants argued their plea on the basis of the existence "of a general requirement of transparency, stemming from Article 1 TEU, Article 11(2) TEU and Article 15 TFEU".[337] In that regard, the General Court has, however, held that "no automatic priority can be conferred on the objective of transparency over the right to protection to personal data".[338] The CJEU has not elaborated further any detail on the principle of transparency and data protection, but it decided on the annulment of EFSA's decision against the disclosure of the information at issue and set aside the General Court's decision. The judgment hence concluded that the public interest in the disclosure of the information outweighed the reasons for keeping part of the document at issue confidential. At a closer look, in *ClientEarth and PAN Europe v EFSA* next to transparency as a risk assessment feature entailing EFSA's decision making process it surfaces the "typical" openness linked to access to documents.[339] In this vein, it has been argued that in the CJEU reasoning is striking the "apparent irrelevance of considerations aimed at balancing personal data protection with the seemingly pervasive right of the public not to trust professionals involved in the Authority's scientific decision-making".[340] The Court has indeed held that the "the mere suspicion of

González Vaqué (2015); Oster (2016), pp. 272–277; Giulimondi (2015), pp. 171–180.

[336] Regulation (EC) No 45/2001 on the protection of individuals with regard to the processing of personal data by the Community institutions and bodies and on the free movement of such data, OJ L 8, 12.01.2001. See Opinion of Mr. Advocate General Cruz Villalón delivered on 14 April 2015, *ClientEarth and Pesticide Action Network Europe (PAN Europe) v European Food Safety Authority (EFSA),* C-615/13 P, ECLI:EU:C:2015:219, esp. para. 54–55.

[337] Judgment of the Court (Second Chamber) of 16 July 2015, *ClientEarth and Pesticide Action Network Europe (PAN Europe) v European Food Safety Authority*, para. 50.

[338] Judgment of the Court (Second Chamber) of 16 July 2015, *ClientEarth and Pesticide Action Network Europe (PAN Europe) v European Food Safety Authority*, para. 51.

[339] For some views on this judgment, see Chap. 5, para. 5.2.2.1.

[340] Gabbi et al. (2020), p. 202.

conflicting interest is sufficient to demonstrate a public interest requiring the disclosure of the personal data".[341]

4.7.3.9 Inconsequential Use

We coded as "inconsequential use" 15 segments from case law that mention transparency without any reference on the Court's interpretation and reasoning. They are not, therefore, examined thoroughly in this analysis. However, in all these case law the inconsequentiality of term transparency seems to be a clear confirmation of the vagueness of the principle of transparency.

4.7.4 Concluding Remarks

The approach of the CJEU became clearer by reading the retrieved segments of a chosen code. To get a visual representation of the Court's standpoint, it is worth reporting the *code frequencies* as a chart[342] (see Fig. 4.2).

As the chart shows, it can be noted that the Court refers to transparency mostly with regard to the *risk assessment principle*. As observed in the General Food Law analysis, risk assessment lies at the core of the food safety regulation, and it appears that the Court has relied on it whenever *transparent* decision-making was required. This centrality is confirmed by the fact that the Transparency Regulation has been adopted with the specific aim of increasing *transparency in the risk assessment*. As seen, the purpose of the new rules is to ensure transparency in all studies and information provided by the industries during the risk assessment process. Along with this obligation, EFSA must make the studies public when receiving the corresponding applications.[343]

The Court's approach consistently underscores that the three principles underlying risk analysis, namely *transparency, independence* and *excellence*, constitute the measure of the Authority's legitimacy.[344] Together these principles define the fundamental framework within which EFSA operates and represent the most relevant for the risk assessment functioning. EFSA defines these principles as "values" on which to base the ethics of its actions.[345] They have also been identified as standards of

---

[341] Gabbi et al. (2020), p. 202.

[342] The MAXQDA function "Code Frequencies" allows to see how many segments and documents were coded with selected codes as a table or diagram. We decided to eliminate from the code frequencies' chart the codes related to "legal provisions" and the "inconsequential use" code.

[343] See para 4.2.

[344] See Judgment of the Court of First Instance (Third Chamber) of 11 September 2002, *Pfizer Animal Health SA v Council of the European Union*, para. 168.

[345] Chiti (2017), p. 11.

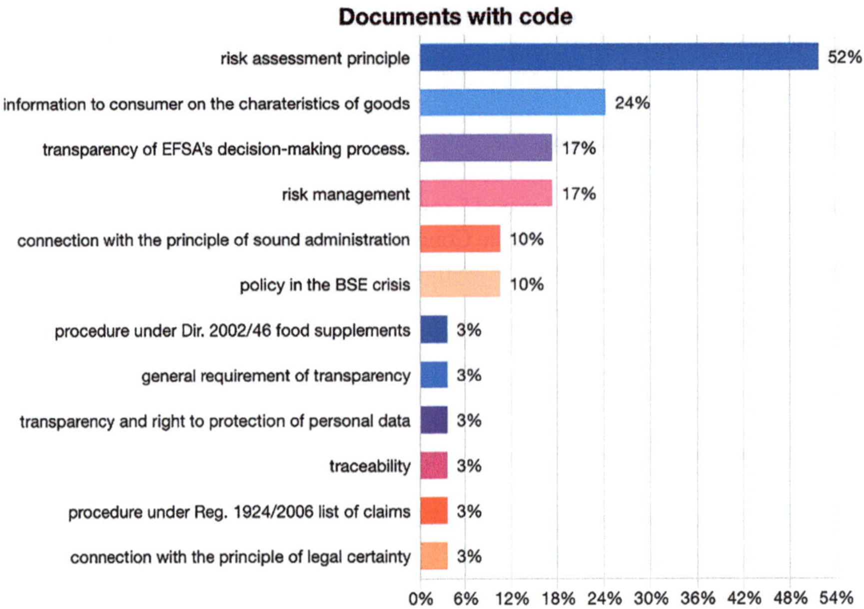

Fig. 4.2 Frequencies of references to transparency in the CJEU case law (MAXQDA 2024)

"good regulatory expertise" by the Commission in the 1997 Communication on consumer health and food safety.[346]

With regard to the legislation, Article 6 (2) General Food Law, as seen, points to transparency, independence and objectivity. The legislator alludes to objectivity instead of excellence as the CFI did in *Pfizer*. However, it has been argued that this mismatch is of limited relevance since the two notions are blurred in the CFI's interpretation and the concept of excellence tends to be confused with the concept of objectivity.[347] In the Court's reasoning, an excellent risk assessment is "as thorough as possible" and carried out "on the basis of the best available scientific data and the most recent results of international research".[348]

It has been argued that from the CJEU perspective the principle of transparency appears relevant not only to risk assessment, but it tends to entail a unified *rationale*

[346] See COM (1997) 183 final, European Commission, "Communication from the Commission on Consumer Health and Food Safety". See also Morvillo (2020), p. 425, who recalled the "good regulatory expertise" principle made by the European Commission in the 2001 White Paper on European Governance and in the 2002 Commission Guidelines on the use of expertise (European Commission, "Communication from the Commission on the Collection and Use of Expertise by the Commission: Principles and Guidelines", COM (2002) 713 final).

[347] Gradoni (2003), p. 202.

[348] Judgment of the Court of First Instance (Third Chamber) of 11 September 2002, *Pfizer Animal Health SA v Council of the European Union*, para. 162. See also Bevilacqua (2017), p. 54.

concerning scientific advice in general.[349] Transparency also serves at correcting the tendency of the advisory process to become self-referential.[350] For instance, transparency calls for carrying out scientific advice that it is understandable and based on reliable information.[351] In the Court's view, transparency functions as a legal requirement of administrative procedures which offers guarantees against risks related to human health. Under this profile, the target of transparency is the public authority.

An additional noteworthy point is the CJEU's deployment of the concept of transparency as related to *information to consumers*. As observed, in a few cases transparency addresses "trading and offer to the public". Accordingly, the Court points out that consumers can raise their awareness about product's characteristics through information on the labels. In this context the Court has recognised the need to find a balance between free movement of goods and the European and national regulations for protecting consumers' interests and public health. Both trade and consumers' interest areas are considered *general principles of food law* in Article 5 and mentioned in the scope of the GFL which "provides the basis for the assurance of a high level of protection of human health and consumers' interest in relation to food [...] whilst ensuring the effective functioning of the internal market".[352] Since the GFL is the foundation of the general part of food law, even specific regulations such as the FIC Regulation or the NHCR take into account both consumers' interest and trade.[353] Under the information to consumers' profile, transparency obligations fall on businesses.

---

[349] Chiti (2017), p. 10. The trend that drew the proliferation of agencies as shaped today came about precisely with the intent of delegating risk assessment responsibilities to them. On point, see Alberti (2018), p. 84.

[350] Chiti (2017), p. 11; Cocconi (2020), p. 69.

[351] Chiti (2017), p. 11. See also Tovo (2016), pp. 249 ff; Bevilacqua (2017), pp. 52–54. According to Fornabaio and Poto (2016), p. 321, the role of the good administrative principles can foster transparency in the risk analysis process. "[...] implementing a core set of principles common to all European administrative proceedings, including the risk analysis procedure, is a priority in progress in the European legislator's agenda: the suggestion to include them in the recitals of the regulation on EU administrative procedures testifies to their growing importance as solid foundations of a common European tradition based on the rule of law and legality" (p. 321).

[352] Article 1 General Food Law.

[353] See, for instance, Article 1 FIC Regulation (Regulation (EU) No 1169/2011 of the European Parliament and of the Council of 25 October 2011 on the provision of food information to consumers, amending Regulations (EC) No 1924/2006 and (EC) No 1925/2006 of the European Parliament and of the Council, and repealing Commission Directive 87/250/EEC, Council Directive 90/496/EEC, Commission Directive 1999/10/EC, Directive 2000/13/EC of the European Parliament and of the Council, Commission Directives 2002/67/EC and 2008/5/EC and Commission Regulation (EC) No 608/2004) and Article 1 NHCR. For an overview of the relevant EU labelling legislation, see Purnhagen and Schebesta (2019). For a study on transparency in food information and labelling in the U.S. Food Law, see Heinzerling (2015), pp. 11–24. At national level, it is worth to note the French law No 2020-699 of 10 June 2020 on the *transparency* of information on agricultural and food products, which established several new sectorial rules in the field of food information. For instance, it provides mandatory original labelling for products made of cocoa and honey, and new rules on food of animal origin names. See also Judgment of the Court (Second Chamber) of 4

The Court has occasionally used transparency with regard to the Commission's policy during the BSE crisis as a risk management's feature.

Finally, from the present analysis it is possible to confirm that transparency is intertwined with the sound administration and legal certainty principles.

Nevertheless, it appears that the CJEU falls short in its judgments of offering an overarching principle of transparency within the food regulatory framework. Apart from the aforementioned case above, the Court often mentions transparency only when it is quoted in articles or recitals of legislative acts.

Accordingly, we can conclude that at the moment the CJEU does not resort to an overarching principle of transparency able to encompass all food law. Nevertheless, it seems that in the Court's view transparency has an instrumental value. On the one hand, transparency is a tool to guarantee a risk assessment procedure that assures, in turn, institutional democracy, legitimacy and accountability. On the other hand, transparency is a guarantee for food safety in both consumers and the market's interests.

4.8 Reflection and Comparison on Transparency as a Principle in EU Law and in the Specific Field of EU Food Law

Finally, it is worth considering the principle of transparency in food law and comparing it with the findings on the principle of transparency in EU law expressed in the previous chapter. Transparency has not been implemented as a general principle of EU law underlying all the possible features that it could have. Nevertheless, the discussion about recognising a general principle of transparency in EU law has centred on access to documents. Only transparency in the form of access to documents has been considered a "serious nominee"[354] to be elevated as a general principle of EU law.

In the General Food Law, the principle of transparency is expressly codified in Articles 9 and 10, while Article 38 establishes the transparency of the Authority's action. Several transparency obligations nonetheless arise in other provisions, especially in conjunction with businesses' responsibilities. As observed in the GFL's analysis, the transparency dimensions we described appear in a considerable number of legal provisions.[355] From a food law perspective, acknowledging that the European food system's principal aim is to keep unsafe food off the market underscores the importance of transparency.[356] In this vein, there is a basis for inferring

October 2024, *Protéines France and Others v Ministre de l'Économie, des Finances et de la Souveraineté industrielle et numérique*, C-438/23, ECLI:EU:C:2024:826.

[354] Prechal and de Leeuw (2007), p. 61. See Chap. 3, para. 3.1 and para. 3.4.

[355] See, para. 4.5.

[356] Bolognini (2013), p. 163.

that transparency may constitute a "golden thread" of food law, as suggested in the White Paper on Food Safety. Despite this perspective, the CJEU devotes little attention to transparency, as the judicial reasoning in the analysed case law does not appear to contribute to the development of an overarching principle. In our opinion, the integrated approach adopted by the European legislator might be undermined if the interpretation of legal provisions fails to connect them and reason in terms of principles and general rules.

Finally, in both EU law in general, and EU food law specifically, the CJEU has linked transparency to other well-established principles of law. In this context, it seems that transparency does not function like a self-standing principle, but it relies on other principles.[357]

In our view, the Court's approach yields a limited comprehension of transparency's potential as an EU food law legal principle. Considering the purposes of transparency found in the legislative perspective, it appears that many of them are not explored by the jurisprudence. At the moment, therefore, it seems that the potential of transparency struggles with its narrow understanding.

References

Alberti J (2018) Le Agenzie Dell'Unione Europea. Giuffré, Milano

Albisinni F (2015) Transparency, crisis and innovation in EU food law. Rivista Della Regolazione Dei Mercati 1:97–115

Alemanno A, Gabbi S (2016) Foundations of EU food law and policy: ten years of the european food safety authority. Routledge, Abingdon

Arcuri A, Hale Hendlin YH (2019) The chemical anthropocene: glyphosate as a case study of pesticide exposures. King's Law J 30(2):234–253

Astill J, Dara RA, Campbell M et al (2019) Transparency in food supply chains: a review of enabling technology solutions. Trends Food Sci Technol 91:240–247

Banati D, Klaus B (2010) 30 years of the rapid alert system for food and feed. Eur Food Feed Law Rev 5(1):10–21

Bartl A (2015) REFIT of food legislation: an opportunity to discuss implementation and enforcement issues. Eur Food Feed Law Rev 10(2):84–89

Bestagno F (2020) La libera circolazione delle merci. In: Arena A, Bestagno F, Rossolillo G (eds) Mercato unico e libertà di circolazione nell'Unione Europea. Casi e materiali di diritto dell'Unione Europea. Giappichelli, Torino, pp 1–116

Beulens AJM, Broens D-F, Folstar P, Hofstede GJ (2005) Food safety and transparency in food chains and networks relationships and challenges. Food Control 16(6):481–486

Bevilacqua D (2017) Democratizzare la tecnica? La partecipazione alle decisioni degli esperti. In: Ziller J, Bassi N (eds) La formazione procedimentale della conoscenza scientifica ufficiale: il caso dell'Agenzia Europea per la Sicurezza Alimentare (EFSA) Giappichelli, Torino, pp 45–70

Bolognini S (2003) Commento all'articolo 55 del reg. (CE) n. 178/2002. La sicurezza alimentare nell'Unione Europea. Commentario a cura dell'IDAIC al reg. (CE) n. 178/2002. Le nuove leggi civili commentate 1–2:443–454

[357] See para. 4.7.3.2. and Chap. 3, para. 3.2.

Bolognini S (2013) Food safety and public communication on food (and feed) safety and risk: the concept of food deemed to be unsafe under review by the court of justice. J Eur Consum Mark Law 2(3):156–169
Borghi P (2003a) Commento all'articolo 5 del reg. (CE) n. 178/2002. La sicurezza alimentare nell'Unione Europea. Commentario a cura dell'IDAIC al reg. (CE) n. 178/2002. Le nuove leggi civili commentate 1–2:191–198
Borghi P (2003b) Commento all'articolo 8 del reg. (CE) n. 178/2002. La sicurezza alimentare nell'Unione Europea. Commentario a cura dell'IDAIC al reg. (CE) n. 178/2002. Le nuove leggi civili commentate 1–2:217–223
Bossis G (2005), La sécurité sanitaire des aliments en droit international et Communautaire. Rapports croisés et perspectives d'harmonisation, Bruylant, Brussels
Bozzini E (2020) Contrasting norms on the use of evidence in risk assessment: the controversy surrounding the carcinogenicity of glyphosate. Health Risk Soc 22(3–4):197–213
Bremmers H, Van der Meulen B, Poppe K, Wijnands J (2008) Administrative burdens in the european food industry: with special attention to the dairy sector. LEI Wageningen. Available via: https://edepot.wur.nl/717. Accessed 8 Jan 2025
Buijze A (2013) The principle of transparency in EU law. Dissertation, Utrecht University
Cane P, Kritzer HM (2010) The Oxford handbook of empirical legal research. Oxford Handbooks in Law, Oxford
Capelli F, Giangiobbe G (2021) La trasparenza nell'analisi del rischio all'interno della filiera alimentare come principio di democrazia. Alimenta 1:1–32
Chamon M (2011) EU agencies between Meroni and Romano or the Devil and the Deep Blue Sea. Common Mark Law Rev 48(4):1055–1075
Chamon M (2023) Op-Ed: 'EU Agencies: Schfting Paradigms of EU Administration'. Available via https://eulawlive.com/op-ed-eu-agencies-shifting-paradigms-of-eu-administration-by-merijn-chamon/. Accessed 18 Apr 2025
Chearnaigh B (2021) Piecemeal transparency: an appraisal of regulation (EU) No. 2019/1381 on the transparency and sustainability of the EU Risk Assessment in the Food Chain. Eur J Risk Regul 12(3):699–710
Chiti E (2017) I principi della valutazione del rischio nel settore della sicurezza alimentare. In: Ziller J, Bassi N (eds) La formazione procedimentale della conoscenza scientifica ufficiale: il caso dell'Agenzia Europea per la Sicurezza Alimentare (EFSA) Giappichelli, Torino pp 9–19
Clark HL (1988) The free movement of goods and regulation for public health and consumer protection in the EEC: the West German Beer Purity Case Note. Virginia J Int Law 28:753–782
Cocconi M (2020) Il ruolo della tecnica nella formazione procedimentale delle decisioni pubbliche. Il caso dell'autorità europea per la sicurezza alimentare. Rivista di diritto alimentare 2:58–71
Conte-Salinas N, Wallau R (2016) The concepts of transparency and openness in European food law. In: Steier G, Patel KK (eds) International food law and policy. Springer, Cham, pp 581–606
Council of Europe/ European Court of Human Rights (2025) Guide on Article 8 of the European Convention of Human Rights—Right to respect for private and family life, home and correspondence. Available at https://www.echr.coe.int/documents/guide_art_8_eng.pdf. Accessed 18 Apr 2025
Craig P, De Búrca G (2020) EU law: text, cases, and materials, 7th edn. Oxford University Press, Oxford
D'Addezio M, Maccioni G (2016) The rapid alert system. In: Costato L, Albisinni F (eds) European and global food law. Cedam-Wolters Kluwer, Milano, pp 301–308
De Boer A (2019a) Transparency and consumer trust in scientific assessments under European food law. In: Urazbaeva A, Szajkowska A, Wernaart B, Tilkin Franssens N, Spirovska Vaskoska R (eds) The functional field of food law. Reconciling the market and human rights. Wageningen Academic Publishers, Wageningen, pp 245–255
De Boer A (2019b) Scientific assessment in European food law: making it future proof. Regul Toxicol Pharmacol 108. https://doi.org/10.1016/j.yrtph.2019.104437

De Vido S (2020) Science, precautionary principle and the law in two recent judgments of the Court of Justice of the European Union on glyphosate and hunting management. DPCE Online 43(2) https://doi.org/10.57660/dpceonline.2020.964

Di Lauro A (2003) Commento all'articolo 16 del reg. (CE) n. 178/2002. La sicurezza alimentare nell'Unione europea. Commentario a cura dell'IDAIC al reg. (CE) n. 178/2002. Le nuove leggi civili commentate 1–2:250–261

Donati A (2020) The Glyphosate Saga, a further but not a final step: the CJEU confirms the validity of the regulation on plant protection products in light of the precautionary principle. Eur J Risk Regul 11(1):148–154

Douma WT (2003) Fleshing out the precautionary principle by the court of first instance. J Environ Law 15:372–405

EFSA (2015) Conclusion on the peer review of the pesticide risk assessment of the active substance glyphosate. EFSA J 13(11). https://doi.org/10.2903/j.efsa.2015.4302

EFSA (2016) EFSA Stakeholder Engagement Approach. Available at https://www.efsa.europa.eu/sites/default/files/EFSA%20Stakeholder%20engagement%20approach_FINAL.pdf. Accessed 18 Apr 2025

EFSA et al. (2023) Peer review of the pesticide risk assessment of the active substance glyphosate. Available via https://efsa.onlinelibrary.wiley.com/doi/epdf/10.2903/j.efsa.2023.8164. Accessed 18 Apr 2025

FAO and WHO (2023) Codex Alimentarius Commission Procedural Manual. Twenty-eight edition, revised. Rome. https://doi.org/10.4060/cc5042en

Ferrari M (2019) Comunicazione del rischio e comunicazione scientifica: spunti per un'analisi interdisciplinare e comparata. Rivista di diritto alimentare 13(3):62–81

Fornabaio L, Poto M (2016) Science and civic engagement in the food sector. Eur Food Feed Law Rev 11(4):315–323

Fuentes VR (2010) The regulation of food risk communication in Spain and the EU. Eur Food Feed Law Rev 5:204–215

Fuentes VR (2017) The Rapid alert system for food and feed—a critical approach. Eur Food Feed Law Rev 12 (2):121–130

Fuentes VR (2019) Food alerts and RASFF. In: Urazbaeva A, Szajkowska A, Wernaart B, Tilkin Franssens N, Spirovska Vaskoska R (eds) The functional field of food law. Reconciling the market and human rights. European Institute for Food Law series. Wageningen Academic Publishers, Wageningen, pp 277–294

Gabbi S (2007) The interaction between risk assessors and risk managers. Eur Food Feed Law Rev 2(3):126–135

Gabbi S, Wood M, Strauss B (2020) Controlling the European food safety authority. In: Sholten M, Brenninkmeijer A (eds) Controlling EU agencies. Edward Elgar Publishing, Cheltenham and Camberley, pp 196–214

Giulimondi F (2015) Corte di Giustizia dell'Unione europea (Sezione seconda), sentenza 16 luglio 2015, causa C-615/13 P, ClientEarth e Pesticide Action Network Europe (PAN Europe) c. European Food Safety Authority (EFSA). Diritto di accesso e trasparenza: veri baluardi della credibilità delle Istituzioni europee. DPCE Online 24(4):171–180

González Vaqué L (2015) European Court of Justice rules in favour of greater transparency in accessing EFSA data. Available at https://scispace.com/pdf/european-court-of-justice-rules-in-favour-of-greater-aphvj279bv.pdf. Accessed 18 Apr 2025

Gradoni L (2003) Commento all'articolo 6 del reg. (CE) n. 178/2002. La sicurezza alimentare nell'Unione europea. Commentario a cura dell'IDAIC al reg. (CE) n. 178/2002. Le nuove leggi civili commentate 1–2:199–203

Hall MA, Wright R (2008) Systematic content analysis of judicial opinions. Calif Law Rev 96:63–122

Heinzerling L (2015) The varieties and limits of transparency in U.S. food law. Food Drug Law J 70(11):11–24

Hiller B (2018) Report on the Judgements of the CJEU (Fourth Chamber) of 19 January 2017 in Case C-282/15 Queisser Pharma GmbH & Co KG v Germany and CJEU (First Chamber) of 27 April 2017 in Case C-672/15 Noria Distributions SARL. Eur Food Feed Law Rev 13(1):34–38
Huizing Edinger W (2014) Food, safety and the behavioural factor of risk. Eur J Risk Regul 5(4):491–504
Karatzia A (2017) The European Citizens' initiative and the EU institutional balance: on realism and the possibilities of affecting EU lawmaking. Common Mark Law Rev 54(1):177–208
Kocharov A (2009) Data ownership and access rights in the European Food safety authority. Eur Food Feed Law Rev 4(5):335–346
Kocharov A (2010) EFSA and identification of emerging risks. Eur Food Feed Law Rev 5(3):144–155
La Manna M (2013) Controversial aspects of the nutrition and health claims regulation. In: Lupone A, Ricci C, Santini A (eds) The right to safe food towards a global governance. Giappichelli Editore, Torino, pp 313–327
Leone L (2021) EFSA under revision: transparency and sustainability in the food chain. Yearb Eur Law 39:536–568
Maijoor S (2011) Market transparency—does it prevent crisis? Presented at FMA Supervision Conference, Vienna, 29 September 2011. Available via https://www.esma.europa.eu/sites/default/files/library/2015/11/2011_322.pdf. Accessed 8 Jan 2025
Mayring P (2014) Qualitative content analysis: theoretical foundation, basic procedures and software solution. Klagenfurt
Mendes J (2017) Bounded discretion in EU law: a limited judicial paradigm in a changing EU. Mod Law Rev 80(3):443–472
Merten-Lentz K (2019) Impact of health claims regulations on the food industry and future innovation potential. In: Astley S (ed) Health claims and food labelling. Royal Society of Chemistry, London
Meyer AH (2006) Risk analysis in accordance with Article 6, Regulation (EC) No. 178/2002. Eur Food Feed Law Rev 1(3):146–153
Morvillo M (2020) Glyphosate effect: has the glyphosate controversy affected the EU's regulatory epistemology? Eur J Risk Regul 11(3):422–435
Nouwt S, Prins C, van der Meulen B, Lugt M (2003) Position paper: transparency in network chains - legal dimensions. Available via: https://edepot.wur.nl/27266. Accessed 8 Jan 2025
Osiejewicz J (2017) Judicial review of EU legislation as an instrument to ensure consistency of national and EU law. In: Rainer A, Martínez-Estay JI (eds) Rule of law, human rights and judicial control of power: some reflections from national and international law. Springer International Publishing, Cham, pp 361–374
Oster J (2016) Of money, poison and secrets: balancing freedom of information against data protection. Eur Data Prot Law Rev 2:272–277
Parisi S, Barone C, Kumar Sharma R (2016) Chemistry and Food Safety in the EU. The Rapid Alert System for Food and Feed (RASFF). Springer International Publishing, Cham
Paulini S (2020) Fact or Fiction? Case C-616/17 and the Compatibility of the EU Authorisation Procedure for Pesticides with the Precautionary Principle. Eur J Risk Regul 11(3): 481-497
Pichler JW, Kaufmann D (2010) The European citizens' initiative: into new democratic territory. Intersentia, Cambridge
Pisciotta G (2018) Sicurezza alimentare, libera circolazione delle merci e regole di responsabilità. Rivista di Diritto dell'Economia, dei Trasporti e dell'Ambiente 16:25–51
Prechal S, de Leeuw M (2007) Dimensions of transparency: the building blocks for a new legal principle? Rev Eur Adm Law 0:51–62
Pugliese S (2017) Il rischio nel diritto dell'Unione Europea: tra principi di precauzione, proporzionalità e standardizzazione. Cacucci, Bari
Purnhagen K (2013) Beyond threats to health: may consumers' interests in safety trump fundamental freedoms in information on foodstuffs? Reflections on Karl Berger v Freistaat Bayern, Wageningen Working Papers in Law and Governance, https://doi.org/10.2139/ssrn.2276899

Purnhagen K (2014) The virtue of Cassis de Dijon 25 years later—it is not dead, it just smells funny. In: Purnhagen K, Rott P (eds) Varieties of European economic law and regulation: liber amicorum for Hans Micklitz. Springer, Cham, pp 315–342
Purnhagen K (2015) The EU's precautionary principle in food law is an information tool! Eur Bus Law Rev 26(6):903–921
Purnhagen K (2019) EU food law: a very short introduction. In: Dries L, Heijman W, Jongeneel R, Purnhagen K, Wesseler J (eds) EU bioeconomy economics and policies. Palgrave Macmillan, Cham, pp 17–26
Purnhagen K, Schebesta H (2019) Food labelling for consumers—EU law, regulation and policy options. Study requested by the PETI committee. Wageningen University. Available via: https://www.europarl.europa.eu/thinktank/en/document/IPOL_STU(2019)608871
Regan D (2010) An outsider's view of Dassonville and Cassis de Dijon: on interpretation and policy. In: Maduro MP, Azoulai L (eds) The past and the future of EU law: the classics of EU law revisited on the 50th anniversary of the Rome Treaty. Hart Publishing, Oxford, pp 465–473
Röttger-Wirtz S (2020) D.2.3 Glyphosate case study: Contribution to the REconciling sCience, Innovation and Precaution through the Engagement of Stakeholders Project (RECIPES). Horizon 2020 Framework Programme of the European Union under Grant Agreement no. 824665. Available via Maastricht University. https://recipes-project.eu/sites/default/files/2020-10/_CS6_Glyphosate.pdf
Rubino V (2013) Information to consumers and operators' responsibilities in food business. In: Lupone A, Ricci C, Santini A (eds) The right to safe food towards a global governance. Giappichelli Editore, Torino, pp 303–313
Rusconi G (2016) Food safety and policy in the European Union. In: Steier G, Patel KK (eds) International food law and policy. Springer, Cham, pp 451–483
Santini A (2013) European food law ten years after Regulation (EC) No 178/2002. In: Lupone A, Ricci C, Santini A (eds) The right to safe food towards a global governance. Giappichelli, Torino, pp 251–258
Santini A (2014) Il sistema di allarme rapido per gli alimenti e i mangimi: brevi considerazioni alla luce del regolamento della Commissione contenente le disposizioni di applicazione. Scritti in memoria di Maria Rita Saulle, pp 1433–1444
Santini A (2019) L'iniziativa Dei Cittadini Europei. Quale Contributo Alla Legittimità Democratica Dell'Unione? Editoriale Scientifica, Napoli
Schebesta H (2018) Content analysis software in legal research: a proof of concept using ATLAS. Ti. Tilburg Law Rev 23(1–2):23–33
Schebesta H (2020) On legal value of implementing acts at the CJEU case note on Schwabe/Queisser Pharma (C-524/18). J Eur Consum Mark Law 9(6):248–252
Schebesta H, Candel JJL (2020) Game-changing potential of the EU's farm to fork strategy. Nature Food 1:586–588
Schebesta H, Purnhagen K (2024) EU food law. Oxford University Press, Oxford
Scholten M, van Rijsbergen M (2014) The ESMA-short selling case: erecting a new delegation Doctrine in the EU upon the Meroni-Romano Remnants. Legal Iss Eur Integr 41(4):389–405
Simoncini M (2020) Administrative regulation beyond the non-delegation doctrine. A study on EU agencies. Bloomsbury, London
Spirovska Vaskoska R, Van der Meulen B, Van der Velde M (2020) Process: hygiene, traceability and recall. In: Van der Meulen BM, Wernaart B (eds) EU food law handbook. Wageningen, Wageningen Academic Publishers, pp 287–325
Steiner J (2007) The limits of state liability for breach of European community law. Eur Public Law 4:69–109
Szajkowska A (2009) From mutual recognition to mutual scientific opinion? Constitutional framework for risk analysis in EU food safety law. Food Policy 34(6):529–538
Szajkowska A (2011) Different actors, different factors - science and other legitimate factors in the EU and National Food Safety Regulation. Eur J Risk Regul 2(4):523–539

Tosun J (2013) How the EU handles uncertain risks: understanding the role of the precautionary principle. J Eur Public Policy 20(10):1517–1528. https://doi.org/10.1080/13501763.2013.834549
Tosun J, Varone F (2020) Politicizing the use of glyphosate in Europe: comparing policy issue linkage across advocacy organizations and countries. J Comp Policy Anal 23(5–6):607–624
Tovo C (2016) Le agenzie decentrate dell'Unione europea. Editoriale Scientifica, Napoli
Van der Meulen B (2007) Transparency & disclosure: legal dimensions of a strategic discussion. Eur Food Feed Law Rev 2(5):270–278
Van der Meulen B (2012) The core of food law: a critical reflection on the single most important provision in all of EU food law. Eur Food Feed Law Rev 7:117–125
Van der Meulen B (2013) The Structure of European food law. Laws 2:69–98. https://doi.org/10.3390/laws2020069
Van der Meulen B (2020) Food law: development, crisis and transition. In: Van der Meulen BMJ, Wernaart B (eds) EU Food Law handbook. Wageningen, Wageningen Academic Publishers, pp 137–158
Van der Meulen B, Szajkovska A (2020) The general food law: general provisions of food law. In: Van der Meulen BMJ, Wernaart B (eds) EU food law handbook. Wageningen, Wageningen Academic Publishers, pp 167–203
Van Gestel R, Micklitz H-W (2014) Methods in European legal scholarship. Eur Law J 20:292–316
Vandamme T (2006) Joined Cases C-154/04 and C-155/04, Alliance for National Health and Nutri-Link v. Secretary of State for Health and the National Association of Health Stores Health Food Manufacture. Leg Iss Econ Integr 33(3):305–318
Vanuzzo M (2021) The general food law between the market and the 'New Transparency Rules'. Revue européenne de droit de la consommation (R.E.D.C.) 2:189–208
Vapnek J (2007) Legislative implementation of the food chain approach. Vanderbilt J Transnatl Law 40(4):987–1014
VERBI Software (2024) MAXQDA 2024 Online Manual. Available at https://www.maxqda.com/download/manuals/MAX24-Online-Manual-Complete-EN.pdf. Accessed 18 Apr 2025
Viti D (2003) Commento agli artt. 9-10 del reg. (CE) n. 178/2002. La sicurezza alimentare nell'Unione europea. Commentario a cura dell'IDAIC al reg. (CE) n. 178/2002. Le nuove leggi civili commentate (1–2):223–227
Vos E (2004) Antibiotics, the Precautionary Principle and the Court of First Instance: Cases T-13/99 Pfizer Animal Health SA v Council of the European Union and T-70/99 Alpharma Inc. v Council of the European Union. Maastricht J Eur Comp Law 11(2):187–200
Vos E (2013) EU Food Safety Regulation and Trust-enhancing Principles. In: Ellefson W, Zach L, Sullivan D (eds) Improving Import Food Safety. Wiley-Blackwell, Hoboken, 111–133
Vos E, Volpato A, Bellenghi G (2023) Independence and transparency policies of the European Food Safety Authority (EFSA), publication for the Committee on Environment, Public Health and Food Safety (ENVI), Policy Department for Economic, Scientific and Quality of Life Policies, European Parliament, Luxembourg. Available at https://www.europarl.europa.eu/RegData/etudes/STUD/2023/740080/IPOL_STU(2023)740080_EN.pdf. Accessed 7 May 2025
Weatherill S (2014) Why there is no 'Principle of Mutual Recognition' in EU law (and why that matters to consumer lawyers). In: Purnhagen K, Rott P (eds) Varieties of European economic law and regulation: liber amicorum for Hans Micklitz. Springer, Cham, pp 401–418
Weisskircher M (2020) The European citizens' initiative: mobilization strategies and consequences. Polit Stud 68(3):797–815
World Health Organization & Food and Agriculture Organization of the United Nations (2006) Food safety risk analysis: a guide for national food safety authorities Available via https://iris.who.int/handle/10665/43718 Accessed 18 Apr 2025

Chapter 5
How Far Does the Principle of Transparency Constrain the Actions of Private Actors?

Abstract This chapter describes the legal background and the specific rules on confidentiality that arise from the novelties introduced by the General Food Law reform. It further addresses the tension between confidentiality and transparency as a peculiar characteristic of the EU agencies' context, providing some examples. This part examines EFSA's role in handling access requests and how EFSA must arbitrate between conflicting public interests. Lastly, the chapter analyses recent case law of the Court of Justice, EFSA's practical arrangements, and the rules implementing Regulation (EU) 2019/1381 ("Transparency regulation") to draw some conclusions on the obligations imposed on, and the potential burdens faced by, private actors.

5.1 Introduction

After the in-depth analysis conducted both on the legal provisions of the General Food Law as recently amended and on the CJEU's case law dealing with transparency in the food sector, it is worth delving into some aspects related to *transparency* and *confidentiality*. More in detail, this chapter explores *how far the principle of transparency constrains the actions of private actors at the moment*.

As observed, transparency obligations do not fall only on public authorities. Especially in the food law sector, the target of transparency is also—and mainly—the food business operator ("FBO") as "the natural or legal person responsible for ensuring that the requirements of food law are met within the food businesses under their control".[1] In the realm of the Transparency Regulation, operators are applicants for an authorisation procedure carried out under EFSA's remit. Therefore, they have a precise interest in a positive outcome of substances or products' approval.

[1] See Chap. 2, para. 2.1.3., Chap. 4, para. 4.2.4.

M. Vanuzzo, *Transparency in EU Law*, Studies in European Economic Law and Regulation 29, https://doi.org/10.1007/978-3-032-02345-2_5

The applicant is also the interlocutor of EFSA and the Commission during the course of the procedure.[2]

The following paragraphs take into account the legal background and important judicial pronounces on access to EFSA's documents, which contributed to the discussion on the transparency of risk assessment. Then, they consider the role of the new provisions of the Transparency Regulation. Even if every scrutiny on the issue should be considered as preliminary, some reflections are offered on the new regulatory landscape and its impact on the private sector.

5.2 The Legal Background on Confidentiality

As seen in the previous chapter, the Transparency Regulation was built on the Fitness Check of the General Food Law's findings and the ECI's initiative on "ban glyphosate and protect people and environment from toxic pesticides".[3] The public distrust in this active substance and the case law of the CJEU contributed to enlivening the discussion on transparency in EU food law.

However, to analyse the potential constraint of the new transparency rules on FBOs, we should briefly recall the distinction between *active* and *passive transparency* already described in Chap. 3. Active transparency allows the public to access information about institutions through the initiative of these organs, while in passive transparency the information is directly requested by the public.[4]

Intending to find a solution to the lack of transparency in the risk assessment process, the Transparency Regulation has modified the rules on transparency and confidentiality laid down in Articles 38 and 39 General Food Law.[5] Article 38 requires that all scientific data, studies and other information supporting authorisation applications shall be made public by EFSA. As a consequence, stakeholders and the public in general should have access to important safety-related information assessed by the Authority.

Undoubtedly, this is an example of *active transparency* deriving from the transparency obligation under Article 38 General Food Law.[6] The ultimate goal of this obligation is indeed the disclosure of information, in the light of an enhanced open and regular dialogue among the Authority and the civil society.[7] Moreover, our analysis shows that Article 38 matches several dimensions of transparency. It enhances *democracy* by disseminating information. Disclosure also has an impact on *legiti-*

---

[2] On a reconstruction of procedures pertaining to EU marketing authorisation in light of the conception of "participation rights", see Mendes (2011).

[3] See Chap. 4, para. 4.3.

[4] See Chap. 3, para 3.1.

[5] See more in detail Chap. 3, para. 3.3.3. and Chap. 4, para. 4.4.3. In this Chapter, we refer to "General Food Law" as amended by the Transparency Regulation.

[6] As we argued in Chap. 4, para. 4.5.1.

[7] See Chap. 4, para. 4.5.1.

macy because making the information available has an impact on decision-making procedures by improving its public acceptance. It indeed legitimises the public authorities' functions in the eyes of citizens. Pursuing the scope to reduce information asymmetry, Article 38 further improves *accountability.*[8]

On the contrary, Article 41 General Food Law is a tool of *passive transparency* because it allows requests for access to documents under Regulation (EC) No 1049/2001 and Regulation (EC) No 1367/2001.[9]

The aforementioned distinction between active and passive transparency is also taken into account by the *Practical Arrangement for implementing Regulation on access to documents.*[10] Whereas 17 of this document expressly states that "the *distinction* between the *reactive* transparency obligations of EFSA in terms of responding to an application for public access to documents and the *proactive* transparency obligations of EFSA laid down in Article 38 of Regulation (EC) No 178/2002, as amended by Regulation (EU) 2019/1381, should be taken into consideration". It further sets out that "the evolving case law on common concepts of the legal framework regarding *access to environmental information* should be taken into account when handling applications for access to documents submitted under this Decision".[11]

Given these preliminary considerations, it is worth taking a step back to the broader EU law framework and having a look at the EU law rules on access and confidentiality.

In this context, it is necessary to state that EU agencies are *pivotal actors* in the "operationalisation" of the EU regime of public access to documents.[12] Furthermore, since the authorisation procedure of regulated products requires a prior application by FBOs, the tension between confidentiality and transparency is a peculiar characteristic of EU agencies' context.[13] This dynamic occurs in the framework of regulatory applications concerning food, chemical substances and medical products. Therefore, EU agencies hold information which is at the same time "potentially commercially sensitive" and "highly relevant" for both the EU regulators and the stakeholders and the public at large.[14] Data held by agencies might be of interest to a third country regulator, or in the context of marketing authorisation procedures in third countries outside Europe.[15]

Although it would be interesting to compare confidentiality rules among EU agencies involved in those procedures, such as the European Medicines Agency

---

[8] See Chap. 4, para. 4.5.2.

[9] See Chap. 3 para. 3.3.3. and Chap. 4, para. 4.4.3.

[10] Decision of the Management Board laying down practical arrangements for implementing Regulation (EC) No 1049/2001 and Articles 6 and 7 of Regulation (EC) No 1367/2006 (mentioned also in Chap. 3 para. 3.3.3. and Chap. 4, para. 4.4.3).

[11] Whereas 17 Practical arrangements for implementing Regulation (EC) No 1049/2001.

[12] Korkea-Aho and Leino (2017), p. 1062.

[13] Korkea-Aho and Leino (2017), p. 1062.

[14] Korkea-Aho and Leino (2017), p. 1060.

[15] Westergrent (2016), p. 901; Korkea-Aho and Leino (2017), p. 1060.

(EMA) and European Chemicals Agency (ECHA),[16] in this book the analysis focuses on EFSA's context to grasp some peculiarities of the food safety sector from the private actors' standpoint. EFSA receives a huge number of requests for access to documents. Many of them usually come from the competitors, and applicant companies fear that the disclosure might cause unfair advantages at their expense and breach their inventions.[17]

In handling access requests, EFSA must arbitrate between *conflicting public interests*. Since confidentiality on one hand and transparency on the other are both "key values underpinning risk assessment and risk management", the balance between them is pivotal for a scientific authority.[18] Interestingly, as seen above, new active transparency obligations also fall on EFSA's remit. These rules aim to "*proactively* make information public" in order to foster the interested parties' possibilities to *influence* application procedures.[19] It has been observed that the new provisions requiring the disclosure of all information, data, studies and documents submitted by FBOs to obtain the authorisation or renewal of their products have brought "the degree of proactive transparency to *new levels*, not yet reached by any other EU agency".[20]

Given that information gathering is one of EFSA's main core tasks, information seems to constitute a *tool* through which the Agency governs.[21] As it has been argued, in this field "information is a critically important topic when evaluating the legitimacy and effectiveness of EU agency governance".[22] Since the *Meroni* judgment, agencies' activities are bound by strict review and supervision in the light of objective criteria.[23] However, the evolving complex regulatory framework has led many agencies to assess not only technical matters but also politically contested issues. In the *Short selling* case, the Court concluded that agencies can be delegated powers "that require reconciling public interests" as long as "the underlying legislative framework is sufficiently detailed, and the measures adopted by the EU agencies are subject to judicial review".[24] The exercise of the powers to make relevant political, economic and social decisions requires agencies to be accountable, and

---

[16] Korkea-Aho and Leino (2017) analyse recent debate on ECHA, EFSA and EMA to support the idea of a new "ownership" paradigm upon EU Agencies. For a comparison of the new transparency rules with EMA's framework, see more in detail Chatzopoulou et al. (2020). About EMA, see Westergrent (2016). See also Hickey and Weimer (2022), p. 693 ff.

[17] Autio et al. (2023), p. 188.

[18] Autio et al. (2023), p. 188.

[19] Neusch et al. (2021), p. 188.

[20] Gabbi et al. (2020), p. 209.

[21] See Chap. 3.

[22] See Korkea-Aho and Leino (2017), p. 1065 (and the reference to Judgment of the Court (Grand Chamber), 22 January 2014, *United Kingdom of Great Britain and Northern Ireland v European Parliament and Council of the European Union*, C-270/12, ECLI:EU:C:2014:18).

[23] We already mentioned the *Meroni* judgment in Chap. 3, para. 3.2.1. and Chap. 4, para. 4.2.4.

[24] Korkea-Aho and Leino (2017), p. 1066. See also Chap. 4, para. 4.2.4.

particularly "judicially accountable".[25] In cases of scientific complex issues, as already noted with regards to the *Pfizer* case, the degree of judicial scrutiny is narrow because the agencies benefit from considerable discretion.[26] What seems to prevail in discourses about agencies is their *accountability*, which in our analysis constitutes a transparency dimension.[27]

5.2.1 More in Detail: Provisions Intended to Protect Confidentiality

5.2.1.1 Confidentiality and Access in EU law Provisions and in the Recent Case Law

As described in Chap. 3, Regulation (EC) No 1049/2001 ("Access Regulation") and Regulation (EC) No 1367/2006 ("Aarhus Regulation") set out the legal framework on access to documents.[28] In addition, sectorial legislation lays down specific provisions intended to protect confidentiality.

Access to documents is excluded in many circumstances.[29] The exceptions' regime is of particular importance when analysing the constraints that transparency causes on *private sector*.

As far as Article 4 (2) Access Regulation is concerned, the "*commercial interest*" *exception* leads to the refusal of access, unless there is an *overriding public interest in disclosure*. In interpreting this exception, the CJEU has held that not all information concerning a company or its business relations fall under it.[30] For instance, in case *Terezakis* the Court has specified that "if *all information relating to a company and its business relations* were regarded as being covered by the protection which

---

[25] Volpato (2019), p. 6. The author recalls the judgment *Les Verts* (Judgment of the Court of 23 April 1986, *Parti écologiste "Les Verts" v European Parliament*, Case 294/83, ECLI:EU:C:1986:166) in which the CJEU emphasises that "the EU is a Community based on the rule of law, inasmuch as neither its Member States nor its institutions can avoid a review of the question whether the measures adopted by them are in conformity with the basic constitutional charter, the Treaty".

[26] Gabbi et al. (2020), p. 205. As observed, in the case *Artegodan* (Judgment of the Court of First Instance (Second Chamber, extended composition) of 26 November 2002, *Artegodan GmbH and Others v Commission of the European Communities,* Joined cases T-74/00, T-76/00, T-83/00, T-84/00, T-85/00, T-132/00, T-137/00 and T-141/00, ECLI:EU:T:2002:283) concerning judicial review of an opinion of a scientific committee, "the Court found that while it could not substitute its own judgement for that of scientific committee, it could review the proper functioning of the committee, the internal consistency of the opinion and the statement of reasons contained therein, and in particular whether there was a comprehensible link between the scientific findings and the conclusions drawn on that basis" (p. 205). The reference to *Pfizer case* is in Chap. 4.

[27] On accountability of EU agencies, see also Vos (2005), pp. 125–129.

[28] See Chap. 3, para. 3.3.3.

[29] See Chap. 3, para. 3.3.3.

[30] Korkea-Aho and Leino (2017), p. 1067.

must be given to commercial interests following Article 4 (2), first indent, of Regulation No 1049/2001, *effect would not be given to the general principle of giving the public the widest possible access to documents held by the institutions*".[31]

More recently, the Court has ruled about EMA's decisions to grant under Regulation (EC) No 1049/2001 access to several documents, namely toxicology reports and a clinical study report submitted by the appellants in the context of their marketing authorisation applications relating to two medicinal products, one for human use (Case C-175/18 P, "*PCT therapeutics*") and the other for veterinary use (Case C-178/18 P, "*MSD Animal Health*").[32]

In deciding whether the Authority's decision to grant access to those reports had hindered the appellants' commercial interest, the Court stated that the existence of a *risk of misuse of the data* contained in the requested document had to be ascertained.[33] A mere unsubstantiated claim without additional details concerning the *nature, purpose and scope* of the data cannot explain *how* access could specifically undermine commercial interests.[34] The same case law further reaffirmed that the agencies could invoke a *general presumption of secre*cy that an EU institution, body, office or agency can apply when receiving an access request. The CJEU stated that it "is however open to that institution, body, office or agency to *base its decisions in that regard on general presumptions* which apply to certain categories of documents, as considerations of a generally similar kind are likely to apply to requests for disclosure relating to documents of the same nature".[35] Nevertheless, the Court has clarified that it is not *required* to base a decision on a general presumption. The Court has thus concluded that the application of a general presumption of confidentiality is "*merely an option*" for the institution, body, agency concerned, which can always examine the documents to verify whether they fall in

[31] Judgment of the Court of First Instance (Third Chamber) of 30 January 2008, *Terezakis v Commission of the European Communities*, T-380/04, ECLI:EU:T:2008:19, para. 93. Emphasis added.

[32] See Judgment of the Court (Fourth Chamber) of 22 January 2020, *PTC Therapeutics International Ltd v European Medicines Agency*, C-175/18 P, ECLI:EU:C:2020:23 and Judgment of the Court (Fourth Chamber) of 22 January 2020, *MSD Animal Health Innovation GmbH and Intervet International BV v European Medicines Agency*. See Court of Justice of the European Union, Press Release No 6/20, Luxembourg, 22 January 2020, Judgments in Case C-175/18 P PTC Therapeutics International Ltd v European Medicines Agency (EMA), and C-178/18 P MSD Animal Health Innovation and Intervet International v European Medicines Agency (EMA) available at https://curia.europa.eu/jcms/upload/docs/application/pdf/2020-01/cp200006en.pdf.

[33] Judgment of the Court (Fourth Chamber) of 22 January 2020, *PTC Therapeutics International Ltd v European Medicines Agency*, para.78 ff.

[34] Judgment of the Court (Fourth Chamber) of 22 January 2020, *PTC Therapeutics International Ltd v European Medicines Agency*, para. 105. See Court of Justice of the European Union, Press Release No 6/20.

[35] Judgment of the Court (Fourth Chamber) of 22 January 2020, *PTC Therapeutics International Ltd v European Medicines Agency*, para. 58, which recalls Judgment of the Court (Grand Chamber) of 4 September 2018, *ClientEarth v European Commission*, Case C-57/16 P, ECLI:EU:C:2018:660, para. 51 and the case-law cited. Emphasis added.

one of the exceptions laid down in Article 4.[36] However, in those circumstances it is incumbent on the institution to establish whether the "general considerations *normally applicable* to a particular type of document" are actually applicable to a *specific document* which it has been asked to disclose.[37]

As a result of those judgments, the CJEU formalised the non-existence of a general presumption of confidentiality of clinical study reports included in the marketing authorisation application dossier. Moreover, the Court also underlines the need to submit in due time the elements on the back of the exception's request. For instance, the explanations concerning in what manner access to documents could specifically undermine the interests protected by the exception. Given the context of the CJEU's ruling and its outcome, it is worth concluding that a general presumption of confidentiality in food authorisation procedures cannot be established.

As regards the *exceptions system*, we must also recall the pivotal *Turco*[38] case. According to the CJEU, an institution must primarily to consider whether the exception can be applied substantively, the *foreseeable and not purely hypothetical* harm, and the potential outweigh of the *public interest* on it.[39] This statement has been subsequently repeated by a multitude of case law. More recently, it has been reiterated the need for a "*sufficiently serious and reasonably foreseeable risk*" such as to justify the application of the exception.[40] The institution or agency has to substantively evaluate whether or not the risk of *concrete harm* outweighs the public interest in disclosure.

Even if the balancing exercise and the consideration of public interest are included in EFSA's remit, the concept of *public interest* for the purpose of balancing is difficult to define.[41] In the case *ClientEarth* (C-57/16),[42] the Court has argued that "it is rather a *lack* of public information and debate which is likely to give rise to doubts as to whether that institution has fulfilled its tasks in a fully independent manner and exclusively in the *general interest*".[43] The case at stake specifically concerns access to documents regarding the impact assessment procedure in environmental matters. The non-profit environmental law organisation ClientEarth

[36] See Judgment of the Court (Fourth Chamber) of 22 January 2020, *PTC Therapeutics International Ltd v European Medicines Agency*, para. 61.

[37] See Opinion of Advocate General Cruz Villallon delivered on 3 October 2013, C-365/12 P *European Commission v EnBW Energie Baden-Württemberg*, p. 30.

[38] Judgment of the Court (Grand Chamber) of 1 July 2008, *Kingdom of Sweden and Maurizio Turco v Council of the European Union*, Joined cases C-39/05 P and C-52/05 P, ECLI:EU:C:2008:374. See also Chap. 3, para. 3.3.3.

[39] See also Judgment of the General Court (Seventh Chamber, Extended Composition) of 22 March 2018, *Emilio De Capitani v European Parliament*, Case T-540/15, ECLI:EU:T:2018:167, para 62.

[40] See, for instance, Judgment of the General Court (Second Chamber) of 21 April 2021, *Laurent Pech v Council of the European Union*, Case T-252/19, ECLI:EU:T:2021:203, para.50.

[41] Curtin and Leino-Sandberg (2016), p. 6.

[42] Judgment of the Court (Grand Chamber) of 4 September 2018, *ClientEarth v European Commission*, Case C-57/16 P.

[43] Judgment of the Court (Grand Chamber) of 4 September 2018, *ClientEarth v European Commission*, Case C-57/16 P, para. 104. Emphasis added.

requested the documents in order to effectively participate in the Commission's decision-making process. As it has been observed, the CJEU refrained from a "principled statement that general presumptions of non-disclosure" cannot be applied to legislative documents, while underscoring the particular importance of transparency in legislative procedures.[44] Even if placed in the context of the early stages of legislative action and precisely in access with regard to impact assessment, *ClientEarth* forms an explicit recognition of the key value of transparency and access to documents as tools for citizens' participation in the decision-making process.[45] In this vein, access to information in good time and at a point that enables citizens to make their views known is at the core of the exercise of democratic rights.[46]

The case *ClientEarth* seems consistent with the previous case law *ClientEarth and PAN Europe c. EFSA*,[47] which emphasises the role of accountability and transparency as tools capable of bringing each institutional subject to its own responsibilities.[48] The concept of information of public interest balances other exceptions to access to documents, as the CJEU states in *ClientEarth and PAN Europe c. EFSA*. When an access request concerns "information of obvious public interest", the balancing exercise is undoubtedly different compared with "when access is sought to documents quite devoid of public interest".[49] In the case under scrutiny, the judgment allowed access to the opinions and curricula of the experts to whom EFSA had turned and to the link between the opinions and their authors to the end of verifying that the experts' decision is equidistant from conflicting interests and devoid of commingling with industrial aims that would seriously affect the decision-making procedures.[50]

As mentioned in Chap. 3, when access to information relates to the environment, the Regulation (EC) No 1367/2006 ("Aarhus Regulation") and Directive 2003/4/EC implementing the Aarhus Convention come into play.[51] Article 6 Aarhus Regulation expressly states that with regard to Article 4 (2) Access Regulation "an overriding public interest in disclosure shall be deemed to exist where the information requested relates to emissions into the environment". Therefore, Article 6 Aarhus Regulation

---

[44] Mendes (2020), p. 24.

[45] Ankersmit (2018).

[46] Ankersmit (2018). See, Hickey and Weimer (2022), p. 705.

[47] Case C-615/13 P, Judgment of the Court (Second Chamber) of 16 July 2015, *ClientEarth, Pesticide Action Network Europe (PAN Europe) v European Food Safety Authority (EFSA)*, ECLI:EU:C:2015:489. See also Chap. 3, para. 3.3.3. and Chap. 4, para. 4.7.3.7 and 4.7.3.8.

[48] Giulimondi (2015), p. 179.

[49] As remarked by Advocate General Cruz Villalon in its opinion delivered on 14 April 2015, Case 615/13 P *ClientEarth, Pesticide Action Network Europe (PAN Europe) v European Food Safety Authority (EFSA)*, ECLI:EU:C:2015:219, para. 54.

[50] According to Gabbi et al. (2020), p. 202, the CJEU reasoning is not entirely convincing since "it appears to be based more on principled statements". Furthermore, the authors argue that "what is striking is the apparent irrelevance of considerations aimed at balancing personal data protection with the seemingly pervasive right of the public not to trust professionals involved in the Authority's scientific decision making".

[51] See Chap. 3, para. 3.3.3.

creates a presumption of the existence of public interest for the Access Regulation scope. In this respect, relevant case law of the CJEU clarifies that the concept of information as linked to emissions to environment covers also information concerning the "nature, composition, quantity, date and place of the actual or *foreseeable* emissions"[52] of those products or substances they refer to, as well as the "data concerning the *medium to long-term consequences* of those emissions on the environment"[53] such as the residues in the environment. More in detail, the CJEU has held that the term "emissions into the environment" could not be interpreted restrictively and information resulting from the use of plant protection products on plants or soil is relevant to environmental protection as information on emissions emanating from sources of industrial installations.[54] Accordingly, the Court has declared that information on emissions into the environment covers also studies which seek to establish the "*toxicity, effects and other aspects of products or substances* under the most unfavourable realistic conditions which could possibly occur", and studies carried out "in conditions as close as possible to normal agricultural practice and conditions which prevail in the area where the product or

[52] Judgment of the Court (Fifth Chamber) of 23 November 2016, *European Commission v Stichting Greenpeace Nederland and Pesticide Action Network Europe (PAN Europe)*, C-673/13 P, ECLI:EU:C:2016:889, para. 79. Emphasis added. In Case C-673/13, the associations Stichting Greenpeace Nederland and PAN Europe submitted a request to the Commission, under Regulation (EC) No 1367/2006 on the application of the Aarhus Convention's provisions concerning access, to disclose documents related to the glyphosate's marketing authorisation. The Commission granted access to those documents in part, witholding parts of the draft assessment report prepared by the Rapporteur Member State (Germany). It justified its refusal by arguing that the material in question contained confidential information on intellectual property rights of companies applying for the glyphosate authorisation, and the two associations brought an appeal before the EU General Court. See van Maldegem (2016); Buonsante and Friel (2017).

[53] Judgment of the Court (Fifth Chamber) of 23 November 2016, *Bayer CropScience SA-NV and Stichting De Bijenstichting v College voor de toelating van gewasbeschermingsmiddelen en biociden*, C-442/14, ECLI:EU:C:2016:890, para 96. In Case C-442/14, Bijenstichting, a bee-protection association in the Netherlands, submitted to the Dutch plant protection and biocidal products authority (Ctgb) a request for access to 84 documents on marketing authorisations of certain plant protection products (PPPs). Bayer was holding a large number of those authorisations. Therefore, it opposed to that disclosure arguing that it would breach its copyrights and adversely affect the "confidentiality of commercial or industrial information". In 2013, the Authority nevertheless granted access to 35 of the 84 requested documents, on the basis that they contained "information on emissions into the environment" which under Directive 2003/4/EC on public access to environmental information, must be available even if such a disclosure could negatively affect the confidentiality of commercial or industrial information. This directive makes clear that commercial and industrial confidentiality may not be invoked to forbid the disclosure of such information. Both Bijenstichting and Bayer appealed the Dutch Authority's decision before a national court, which referred the case to the CJEU for a preliminary ruling. See van Maldegem (2016); Buonsante and Friel (2017).

[54] Buonsante and Friel (2017), p. 455. The authors highlight that the CJEU pointed out that no distinction should be made between the concepts of emission, releases and discharges, as the Aarhus Convention in Article 4(4)(d) merely refers to "information on emissions, which is relevant for the protection of the environment".

substance is to be used".[55] Interestingly, the CJEU has interpreted the Aarhus Regulation by establishing a presumption that the public interest in obtaining information prevails over commercial interests.

In the peculiar context of Plant Protection Products—Regulation (EC) 1107/2009, also called Regulation on plant protection products or "PPPs"[56] amended by the Transparency Regulation—two main judgments involve EFSA's responses to access requests. Precisely, the cases *Tweedale*[57] and *Hautala*[58] both concern request of access to documents related to glyphosate's toxicity and carcinogenicity studies and similar complaints against EFSA for denying access to industry studies. The General Court has further confirmed that Article 6 Aarhus Regulation establishes an absolute presumption of the existence of an overriding public interest in disclosure and "it derogates from the rule requiring the weighing up of the interests laid down in Art. 4 (2), first indent of Regulation (EC) No 1049/2001".[59] Furthermore, according to the CJEU, for the presumption under Article 6 Aarhus Regulation it is crucial to define the intensity of the link between the requested information and the emissions into the environment.[60] A link that is too vague would deprive the commercial interests' exception, compromising "the balance which the EU legislature intended to maintain between *the objective of transparency* and the protection of those interests" culminating "in disproportionate interference with the protection of business secrecy".[61] The General Court found that the studies of the applicant (business operator) to which EFSA refused access must be regarded as constituting information that relates to emissions into the environment and that an overriding public interest in disclosing the studies were deemed to exist. Thereby, the Court ordered disclosure of the requested studies. In *Tweedale* and *Hautala*, the CJEU for the first time qualified toxicity and carcinogenicity studies in a PPPs dossier as information related to the environment and eliminated the need to balance the existence of an overriding public interest against the protection of commercial interests.[62] On this point, it has been argued that beyond the victories for the individual claimants, one

[55] Judgment of the Court (Fifth Chamber) of 23 November 2016, *Bayer CropScience SA-NV and Stichting De Bijenstichting v College voor de toelating van gewasbeschermingsmiddelen en biociden*, para. 91. Emphasis added. For some critical remarks about the definition with regard toxicity studies see Buonsante and Friel (2017), p. 457.

[56] See Chap. 4, para. 4.3.

[57] Judgment of the General Court (Eighth Chamber) of 7 March 2019, *Antony C. Tweedale v European Food Safety Authority*, Case T-716/14, ECLI:EU:T:2019:141.

[58] Judgment of the General Court (Eighth Chamber) of 7 March 2019, *Heidi Hautala and Others v European Food Safety Authority*, Case T-329/17, ECLI:EU:T:2019:142.

[59] Judgment of the General Court (Eighth Chamber) of 7 March 2019, *Antony C. Tweedale v European Food Safety Authority*, para. 57. See also Morvillo (2019), p. 423 and Donati (2020).

[60] Morvillo (2019), p. 423.

[61] See Morvillo (2019), p. 423, who recalls Judgment of the General Court (Eighth Chamber) of 7 March 2019, *Antony C. Tweedale v European Food Safety Authority*, para. 89 and Judgment of the General Court (Eighth Chamber) of 7 March 2019, *Heidi Hautala and Others v European Food Safety Authority*, para. 96.

[62] Morvillo (2019), p. 424.

important implication of the Court's ruling is that it "brings regulatory science back into the domain of public debate".[63]

The described judicial line of case law is relevant not only for the issues involved, such as the glyphosate episode and the pitfalls concerning endocrine-distruptors substances,[64] but it also frames the idea that all confidential substance-related *information* linked to commercial interests should be disclosed when considered to be related to emissions into the environment. Consequently, it has an impact on the possibility of claiming confidentiality of documents applied in regulatory procedures. Following those judgments on the core concepts of access to environmental information, interest in information held by EFSA and other EU agencies should increase.[65] In this vein, businesses are concerned that the data they must provide under the access legal framework might become available to their competitors. In the agri-food sector where are involved on the one hand consumer's safety, and commercially sensitive information on the other hand, FBOs concerns and conflicting interests constitute a matter of discussion.

It should also be recalled that the companies' information provided to EFSA for marketing authorisation, registration or approval is third-party property. As mentioned, the Access Regulation provides a consultation procedure when requested documents are "third-party documents", meaning a natural or a legal person or any entity outside the institution concerned, including Member States, other EU or non-EU institutions and bodies and third countries.[66] The CJEU stated in *IFAW* that the consultation of the third party is generally a precondition for determining whether the exceptions to the right of access provided for in Article 4 (1) and (2) of Regulation (EC) No 1049/2001 are applicable in the case of third-party documents.[67] Therefore, the agency should assess the exceptions' applicability following the consultation of a third party.

5.2.1.2 Confidentiality and Access in the General Food Law. EFSA Internal Rules and the Implementation of the Transparency Regulation

In practice, agencies such as EFSA apply the Regulation (EC) No 1049/2001 on the basis of their founding regulations. As previously described in Chap. 4, the Transparency Regulation has modified the rules on transparency and confidentiality laid down in Articles 38 and 39 General Food Law.[68] These rules are of fundamental

[63] Morvillo (2019), p. 426. See also Arcuri and Hale Hendlin (2019), p. 247.

[64] Judgment of the General Court (Third Chamber) of 20 September 2016, *Pesticide Action Network Europe (PAN Europe) v European Commission*, Case T-51/15, ECLI:EU:T:2016:519.

[65] As also observed by Korkea-Aho and Leino (2017), p. 1063.

[66] Article 3(b) Regulation (EC) No 1049/2001. See Chap. 3, para. 3.3.3.

[67] See Korkea-Aho and Leino (2017), p. 1068.

[68] See Chap. 4, para. 4.4.3.

importance for investigating the potential constraints on private actors. FBOs are, indeed, the "applicants" contributing to EFSA's calls for data or submitting scientific data in marketing authorisations procedures.[69]

Much has changed since the prior regime. Under the previous legal framework, EFSA had implemented rules on public access and transparency with a Decision of its Management Board of 2005.[70] However, the earlier legal note on "Openness, transparency and confidentiality – general principles"[71] does not specify how EFSA takes into account the confidentiality of scientific studies on which it bases its opinions or evaluates confidentiality requests.[72]

In 2014, EFSA published a discussion paper expressing the aim to transform the Authority into an "Open EFSA" confirming its commitment to balance data owners' interests and the public's expectations.[73] Moreover, EFSA declared that "transparency and openness should therefore be result driven and not a goal *per se*, to ensure that EU citizens obtain value for the money invested in the EU project. Furthermore, greater involvement and participation could also hide potential risks, such as disproportionate influence of a limited number of actors or loss of control by the Authority over the content of a document".[74]

In marketing authorisation procedures, Article 38 and Article 39 General Food Law represent the basic rules. As the CJEU has pointed out, the principle of

---

[69] According to Article 3 Practical Arrangements on transparency and confidentiality (Decision of the Executive Director of the European Food Safety Authority Laying down practical arrangements concerning transparency and confidentiality of. 18.01.2021 available at https://www.efsa.europa.eu/sites/default/files/corporate_publications/files/210111-PAs-transparency-and-confidentiality.pdf) applicants means "(i) any natural or legal person submitting an application or notification under Union law; (ii) any natural or legal person submitting scientific data and information for evaluation to the Authority pursuant to established sectoral Union law procedures; (iii) where permitted under sectoral Union law procedures and/or in the absence thereof, any natural or legal person submitting information voluntarily to the Authority upon which the Authority is expected to base its scientific outputs within the meaning of Article 38(1)(d) of the General Food Law Regulation; (iv) any natural or legal person who has produced information supporting a request from the European Parliament, the Commission and the Member States for a scientific output and therefore having a direct interest with respect to the closed list of information items for which confidentiality treatment can be requested as laid down in the Annex". As expressed in EFSA (2021a) Q&A document: "the Practical Arrangements in question adopt a pragmatic approach of considering as applicants not only persons submitting applications dossiers" (p. 30). In the context of Plant Protection Products, see Decision of the Executive Director of the European Food Safety Authority Laying down practical arrangements concerning confidentiality in accordance with Articles 7(3) and 16 of Regulation (EC) No 1107/2009 (available at https://www.efsa.europa.eu/sites/default/files/corporate_publications/files/210111-PAs-confidentiality-Artt-7-and-16-of-regulation-1107-2009.pdf).

[70] Decision of the management board of the European Food Safety Authority concerning implementing measures of transparency and confidentiality requirements of 10.03.2005 (repealed).

[71] Openness, transparency and confidentiality 16.09.2003, 13 (available at https://www.efsa.europa.eu/sites/default/files/corporate_publications/files/transparencyprinciples.pdf).

[72] See also Korkea-Aho and Leino (2017), p. 1073.

[73] EFSA (2014).

[74] EFSA (2014), p. 7.

openness and transparency in Article 38 General Food Law forms an *essential pillar* of EFSA's activities and confidentiality represents an exception.[75] Regarding the implementation of the new provisions introduced by the Transparency Regulation, EFSA has published the Practical Arrangements.[76] These are detailed arrangements for how the new rules operate in practice and are provided to "help stakeholders to better understand how the new process and tools affect them and what adjustment they need to make to operate effectively in the new environment".[77] As the EU Commission and EFSA declared, these are considered binding in interpreting the new legislative framework. In this regard, we should take into account not only the *Decision of the Management Board laying down practical arrangements for implementing Regulation (EC) No 1049/2001 and Articles 6 and 7 of Regulation (EC) No 1367/2006*[78] but also the *Practical Arrangement concerning transparency and confidentiality.*[79] Useful tools to understand the complex framework are also the *Questions and Answers on the Practical Arrangements* ("Q&A") document[80] and the *Administrative guidelines for the processing of applications of regulated products.*[81]

Following the recent amendments, Article 38 General Food Law mandates that all scientific data and documentation in support of authorisation or other assessments be made public and available. EFSA is thus required to operate "with a high level of transparency", making public, among others, scientific outputs, underlying data and other materials that form the basis for supporting applications.[82] Unless the applicant requests confidentiality, the publication follows the Authority's evaluation of the dossier validity or admissibility "without delay" (Article 38 (1)).

75 Gabbi et al. (2020), p. 208. They refer to Case T-675/18 R, *Trifolio-M GmbH, Oxon Italia & Mitsui AgriScience International v EFSA*.

76 Article 38 (3) Transparency Regulation expressly provides that "the Authority shall lay down the practical arrangements for implementing the transparency rules referred to in paragraph 1, 1a and 2 of this Article, taking into account Articles 39 to 39g and 41".

77 See https://www.efsa.europa.eu/en/news/new-rules-transparency-detailed-arrangements-stakeholders.

78 See para 2.

79 Decision of the Executive Director of the European Food Safety Authority Laying down practical arrangements concerning transparency and confidentiality.

80 EFSA (2021a).

81 EFSA (2021b).

82 See also Practical Arrangements concerning transparency and confidentiality, Article 5 "Proactive disclosure of information, documents or data in relation to which confidentiality requests may not be submitted" and "Article 6 "Proactive disclosure of information, studies, documents or data in relation to which confidentiality requests may be submitted" where there is the list of items publicly available.

EFSA publishes the items on its OpenEFSA portal,[83] IUCLID[84] or in EFSA's scientific journal.[85] The OpenEFSA portal makes available to the public documents, information or data deemed not confidential by the applicant, the Rapporteur Member State, the Commission or EFSA depending on the legal framework and the different stages of a potential confidentiality request.[86]

Article 7.1 Practical Arrangements concerning transparency and confidentiality provides that the proactive disclosure of information, documents and data "shall not provide permission or license for their dissemination, re-use, reproduction, or exploitation in breach of rules concerning existing rights, including intellectual property rights, or data exclusivity rules under Union law".[87] Therefore, the publication and the related possibilities for downloading, printing and searching of information according to Article 38 (1) do not allow the use of documents in breach of IP rights.[88] The Transparency Regulation has indeed recognised and preserved existing provisions concerning intellectual property rights ("IP Rights") and investments made by innovators ("data exclusivity rules").[89] Namely, Article 38 (1a) clarifies that the IP rights and the protection of documents and innovators' investments remain unaffected.[90] On this point, an ongoing debate has emerged about data ownership and the potential shift in ownership when scientific data are transferred from

[83] OpenEFSA is the single public interface for all information related to EFSA's scientific work, see https://open.efsa.europa.eu/.

[84] IUCLID is a software to record, store, maintain and exchange data on intrinsic and hazard properties of chemical substances, see https://iuclid6.echa.europa.eu/.

[85] The open access online scientific journal that publishes EFSA's scientific outputs, available at https://efsa.onlinelibrary.wiley.com/.

[86] See EFSA (2021a) point 4, p. 32.

[87] Users of the OpenEFSA portal will be allowed to get access to the documents, information or data supporting application dossiers only after having accepted EFSA's terms of reference complying with Article 38(1a) (see EFSA (2021a), p. 32.

[88] Neusch et al. (2021), p. 194. As recalled in Chap. 4, para 4.3, Article 39e on the protection of the personal data requires that certain information be made public (including the name and the address of the applicant and the name of the authors of the studies supporting the request and of "all participants and observers in the meetings of Scientific Committee and of the Scientific Panels"). Article 39f introduced the so-called standard data format that EFSA must implement "to ensure the efficient processing of requests".

[89] For a comprehensive overview of the legal and policy framework on data ownerships and access rights in the European Food Safety Authority, see Kocharov (2009), pp. 335–346. In the Practical Arrangement concerning transparency and confidentiality, Article 3 (1d) defines "data exclusivity rules" as "any provisions set out in sectoral Union law, which aim to protect the investment made by innovators in gathering the information and data supporting relevant applications for authorisations". See also WIPO (2018). On this point, ClientEarth (2018) concludes that the wording of the Transparency Regulation is "highly problematic" and the formulation of Article 38 (1a) "may lead to misinterpretation as offering new grounds to claim information as confidential because commercially sensitive, in addition to the one already listed in Article 39".

[90] As highlighted by Neusch et al. (2021), p. 194, the publication and the associated worldwide possibility of downloading, printing and searching the information in accordance with Article 38 (1) (3) General Food Law "does not constitute permission to use, copy or exploit the documents in breach of these rights".

FBOs to the Authority and then published.[91] However, this specific issue goes beyond the scope of the question about the practical constraints that the Transparency Regulation poses on private actors. In this paragraph, the focus is on the potential balance between transparency and confidentiality, notably between *public and private interests* linked to the use and disclosure of documents. In the context of company dossiers submitted to the Authority, it is necessary to distinguish data protection and confidential treatment of FBOs' information submitted to EFSA.[92] On the one hand, data protection means that the information provided by a company for the purposes of a marketing authorisation "cannot benefit another company".[93] On the other hand, confidential treatment implies that the information "cannot be disseminated to third parties without the prior approval of the originator".[94] As it has been observed, while the companies *own the data* submitted to the agencies under IP and copyright legislation, "the concept of ownership is *irrelevant*, and perhaps even inimical to the interests pursued under the public access regulation".[95]

At this point, it is worth digging deeper into the rules encompassing confidentiality, both in the Transparency Regulation and in the Practical Arrangements and highlighting the provisions mostly connected with the potential burdens on private actors.

Confidentiality is allowed only in a limited number of cases that the current version of Article 39 General Food Law sets out and in cases provided by specific legislation. The provision establishes both the procedure and a horizontal list of categories of information generally regarded as commercially sensitive the disclosure of which could potentially impact the applicants' interests. The confidentiality regime under Article 39 introduces a twofold mechanism for confidentiality requests based on demonstration of the *potential harm to the applicant's interests to a significant degree*, and on the circumstance that the request is grounded on the information included in the categories listed in Article 39 (2) General Food Law.[96]

As regards the procedure, in Chap. 4 it is already mentioned that according to Article 39a General Food Law, when applicants invoke confidentiality, they must provide a "verifiable justification" explaining how the disclosure of specific information would harm their interests "to a significant degree". Additionally, the

---

[91] See Sapienza (2019); Korkea-Aho and Leino (2017).

[92] Kocharov (2009), p. 341.

[93] Kocharov (2009). p. 341. As regards data protection and novel food, see Simpson (2016), pp. 309–314; Holle (2014), pp. 280–284; Monaco (2023), pp. 172–178.

[94] Kocharov (2009), p. 341.

[95] Korkea-Aho and Leino (2017), p. 1064. Emphasis added. See EFSA (2021a), p. 33: the Q&A about the materials disclosed by EFSA and subject to IP rights clarifies that "Applicants may also consider supporting their applications by relying on scientific publications available to the public (e.g. upon payment of a fee). Where applicants may not have or may not be able to obtain the necessary IPRs by the rightful owner (e.g. scientific journals) for the reproduction of those publications on the Open EFSA portal (for public dissemination purposes), they will have the possibility of providing instead the full bibliographic references allowing the correct identification of the publication by the public".

[96] Article 39 (2). See Rusconi and Cozzolino (2021), p. 305.

applicant must submit a non-confidential version excluding the contested parts. The complete version instead must include all information, the confidential parts along with a detailed explanation supporting the confidentiality requests. EFSA then publishes the non-confidential version and issues a decision on the confidentiality request following a concrete and case-specific examination pursuant to Article 39b General Food Law within 10 weeks of the date of receipt. According to Article 39b (2), within two weeks of the notification of the Authority's decision, the applicant may submit a *confirmatory application with suspensive effects* asking EFSA to reconsider its decision. The Authority must decide on this within three weeks.[97]

EFSA is responsible for the decision-making on confidentiality requests whenever scientific data, studies or other information are transmitted for the purposes of scientific output. However, in some processes, the responsibility for confidentiality decision-making is assigned to the Commission[98] or the competent national authority.[99]

As to the *verifiable justification of the harm*, the applicant must provide EFSA with *concrete* evidence demonstrating the harm. Article 10 Practical arrangements concerning transparency and confidentiality lays down the *substantive requirements for confidentiality requests*. Namely, after identifying clearly the relevant parts of the submitted information claimed as confidential,[100] the applicant should explain "comprehensively and in plain language" the reason why the information should be

[97] Pursuant to Article 39b (c), even before the formal decision, EFSA shall inform the applicant of the decision it intends to take and in case of disagreement with the assessment of the Authority, the applicant may state its views or withdraw its application within two weeks of the date on which it was notified of the Authority's position.

[98] EFSA (2021a) Point 9, p. 37 enlists the items whose the Commission is empowered to take confidentiality decisions: "information supporting applications for authorisation of food additives, food enzymes or food flavourings submitted under Regulation (EC) No 1331/2008, where an EFSA opinion is not required in accordance with Article 3(2) thereof; information supporting notifications of the intention to place traditional foods from third countries on the Union markets pursuant to Article 14 of Regulation (EU) 2015/2283; information supporting applications for authorisation of novel foods where an EFSA opinion is not required in accordance with Article 10(3) of Regulation (EU) 2015/2283; information related to the re-evaluation of a food additive, including information submitted under a call for data, considered by EFSA pursuant to Article 4 of Commission Regulation (EU) No 257/2010, where EFSA is not requested to take further steps in accordance with Article 7a(1) thereof".

[99] This is the case for: (i) data and information supporting dossiers for approval of new active substances for use in pesticides pursuant to Article 7 of Regulation EC No 1107/2009; and (ii) information supporting notifications under Directive 2001/18/EC ('GMO Directive'). To ensure consistency in the confidentiality assessment of Member States and EFSA in the field of pesticide active substances, there are some "safeguards", such as "the identical lists of items of information for which confidential treatment may be requested" and Practical Arrangements concerning confidentiality pursuant to Articles 7 and 16 of Regulation EC No 1107/2009, which ensures "the application of comparable procedural and substantive requirements" (Point 9 EFSA (2021a)).

[100] The relevant parts should be "accompanied by a link to and a detailed reference to the exact paragraph, page, line or part thereof as appropriate, where this information is located, in a sufficiently precise manner so as to exclude any information that is not subject to the confidentiality request" (Article 10 (a) Practical Arrangements concerning transparency and confidentiality).

granted confidential status.[101] The explanation shall include a justification as to *why* the *cumulative requirements* listed are deemed satisfied. In particular, the minimum requirements refer to the unavailability of the document in the public domain or its knowledge only by a limited number of persons, the *potential* harm to the applicant's interests and its *significance*, the document's legal acquisition under IP rights, whether the document falls under the definition of "environmental information" (Article 2 Aarhus Regulation), and whether the information concerned has been finalised more than 5 years ("novelty test") before the submission of the confidentiality request (or a demonstration of the potential harm even if the documents were previously finalised).[102]

Article 39 does not define what are the *interests* involved in the confidentiality request. Since Recital 32 Transparency Regulation expressly refers to *commercial interest* and given the overall interpretation of the provisions intended to protect confidentiality, it seems preferable to lean towards this solution.[103] It has been noted that it seems implausible that the disclosure of certain information could harm other interests, such as moral, corporative or personal ones.[104] However, the disclosure of certain information may cause non-material damages, for instance, to *reputation*. In this case, the applicant should demonstrate the existence of a structural or legal obstacle preventing the restoration of their reputation rather than relying on mere speculation, particularly if the applicant can restore it by employing a publicity campaign.[105]

---

[101] Article 10 (b) Practical Arrangements concerning transparency and confidentiality.

[102] The requirements listed in Article 10 (b) Practical Arrangements concerning transparency and confidentiality precisely are: "(i) the document, information or data for which confidential status is requested is not publicly available or is known only to a limited number of persons; (ii) the public disclosure of the document, information or data for which confidential status is requested may potentially harm the interests of the applicant to a significant degree; (iii) explanation or evidence demonstrating to the satisfaction of the Authority that the harm that may be caused is of a significance corresponding at least to 5% of the gross annual turnover for legal persons, or the gross annual earnings for natural persons, for the financial year preceding the calendar year of the submission of the confidentiality request. If the harm is quantified as not reaching this percentage, or the applicant is unable to calculate its impact on their turnover/earnings, the applicant shall provide a specific reason as to why they consider that any public disclosure would potentially harm their interests to a significant degree; (iv) the document, information or data for which confidential treatment is requested is eligible for legal protection and has not been acquired in an unlawful manner; (v) the document, information or data for which confidential status is requested does, or does not, fall under the definition of "environmental information" pursuant to Article 2 of the Aarhus Regulation; (vi) the document, information or data for which confidential status is requested has been finalised in the form in which it was submitted to the Authority up to five (5) years prior to the submission of the confidentiality request. If the document, information or data for which confidential status is requested is older than five (5) years, the applicant shall provide (a) specific reason(s) as to why public disclosure of that information would still potentially harm its interests to a significant degree".

[103] See, Verzijden and Buijs (2020), p. 105.

[104] Rusconi and Cozzolino (2021), p. 306.

[105] Rusconi and Cozzolino (2021), p. 306; De Boer, Morvillo and Röttger-Wirtz (2023), p. 140 ff.

Regarding the element of a *significant degree* of the harm, Article 10 Practical Arrangements requires that *significance* must correspond "at least to 5% of the gross annual turnover for legal persons, or the gross annual earnings for natural persons, for the financial year preceding the calendar year of the submission of the confidentiality request".[106] Whether the harm does not reach this percentage, or the applicant is unable to quantify its impact on their turnover or earnings, the applicant shall provide *a specific reason* which justifies the confidentiality request. For instance, EFSA should take into account various elements, such as the size of the company, the situation in the market, the number of competitors, the economic value of the information and its "secrecy degree", the number of employees and their role in relation to information claimed as confidential.[107] Moreover, the seriousness of the harm can be related to job losses resulting from the departure of qualified staff or the loss of skills and experience, the closure of a factory and subsequent loss of investment, increased production cost, and a fall in sales of products.[108]

As per the items eligible for confidentiality requests, we briefly analyse each of them to better understand the concrete obligations FBOs must comply with. As previously noted, most of these items concern commercial aspects and issues.

Among these, Article 39 (2) (a) General Food Law mentions "the manufacturing or production process", which includes "the method and innovative aspects thereof", as well as *other technical and industrial specifications* inherent to that process or method, except when it is relevant for food safety. This may refer also to specific steps or phases of the entire production or manufacturing process. Even if this is not *innovative*, confidentiality can be granted as long as the applicant demonstrates a *significant degree* of harm to its interests.[109]

The applicant may also demand confidentiality on the basis of the "commercial links between a producer or importer and the applicant or the authorisation holder", in case of a renewal's request (Article 39 (2) (b) General Food Law).

Furthermore, *commercial information* capable of revealing *sourcing, market shares or business strategy* of the applicant can be encompassed in a confidentiality request (Article 39 (2) (c) General Food Law). Since the interpretation of the business strategy can be broad, the applicant might likely support a disputable confidentiality request through this hypothesis.[110]

Lastly, the applicant may cover with confidentiality the "quantitative composition of the subject matter of the request", except when it is relevant for food safety (Article 39 (2) (d) General Food Law). For instance, this can include the confidentiality of the quantitative data on the composition, formulas, major components, proportions of substances in the mixture and so forth.

---

[106] Article 10 (b) (iii) Practical Arrangement concerning transparency and confidentiality.

[107] Rusconi and Cozzolino (2021), p. 306. See EFSA (2021a) Point 13.

[108] Rusconi and Cozzolino (2021), p. 306.

[109] See Rusconi and Cozzolino (2021), p. 307. Vos et al. (2023), p. 61.

[110] Rusconi and Cozzolino (2021), p. 307.

Along with the items listed in the General Food Law, the sectoral legislation amended by the Transparency Regulation sets out specific rules on confidentiality, introducing into the texts additional requirements or specific provisions depending on the piece of legislation.[111] For instance, under Regulation (EU) 2015/2283 on novel foods, confidentiality status may additionally be requested for "information provided in detailed descriptions of starting substances and starting preparations and on how they are used to manufacture the novel food subject to the authorisation", detailed "information on the nature and composition of the specific food*s* or food categories in which the applicant intends to use that novel food", except when it is relevant for safety, and "detailed analytical information on the variability and stability of individual production batches", except when it is relevant for safety (Article 23 (4) Regulation (EU) No 2015/2283).[112] As of genetically modified (GM) food and feed, submission of a new GM food or feed product may include also "DNA sequence information, except for sequences used for the purpose of detection, identification and quantification of the transformation event", and "breeding patterns and strategies" (Article 30 (2) Regulation (EC) No 1829/2003).[113]

Similar provisions are introduced for the other amended legislative acts depending on each regulatory framework.[114]

As regards EFSA internal rules implementing the Transparency Regulation, it should also be mentioned that the handling of the items subject to confidentiality requests have been submitted or granted "confidential status" is under EFSA's remit. The Authority shall ensure that they are made available upon request, to the

[111] See Annex – Items for which confidentiality requests may be submitted, Practical Arrangements concerning transparency and confidentiality.

[112] See also EFSA Administrative guidance for the preparation of applications on novel foods pursuant to Article 10 of Regulation (EU) 2015/2283, available at https://www.efsa.europa.eu/en/supporting/pub/en-6488.

[113] See EFSA (2021c). See also Jennings (2021) p. 172.

[114] For example, within the *feed additive framework*, Article 18(3) of Regulation (EC) No 1831/2003 mentions "the study plan for studies demonstrating the efficacy of a feed additive, specifications of the impurities of the active substance and the relevant methods of analysis developed internally by the applicant, except for impurities that may have adverse effects on animal health, human health, or the environment". The *Food content materials* framework considers "any information provided in detailed descriptions of starting substances and mixtures used to manufacture the substance subject to the authorisation, the composition of mixtures, materials or articles in which the applicant intends to use that substance, the manufacturing methods of those mixtures, materials or articles, impurities, and migration testing results, except for information which is relevant to the assessment of safety", and "the trademark under which the substance shall be marketed as well as the trade name of the mixtures, material or articles in which it shall be used, where applicable" (Article 20 (2) Regulation (EC) No 1935/2004. For *food improvements agents*, Article 12 (3) Regulation (EC) No 1331/2008 adds "information provided in detailed descriptions of starting substances and starting preparations and on how they are used to manufacture the substance subject to the authorisation, and detailed information on the nature and composition of the materials or products in which the applicant intends to use the substance subject to the authorisation, except for information which is relevant to the assessment of safety" and "detailed analytical information on the variability and stability of individual production batches of the substance subject to the authorisation, except for information which is relevant to the assessment of safety".

Commission and the Member States. Moreover, EFSA shall make available internally these items only to those who need to have access to them in order to perform the tasks assigned in the context of their operational responsibilities.[115] Hence, EFSA, the Commission and the Member States shall ensure that items are not disclosed to *unauthorized third parties*. However, the disclosure becomes necessary when urgent action is essential for the purposes of protecting human health, animal health or the environment, pursuant to Article 39 (4)(a) General Food Law. The provision of such a "transparency safeguard" requires EFSA to verify such urgent reasons before disclosing information previously qualified as confidential.

The release of commercially sensitive information may be challenged before the CJEU or the European Ombudsman. Pursuant to Article 12 Practical arrangements on transparency and confidentiality EFSA's decisions can be challenged by lodging an action contesting their legality before the CJEU according to Article 263 TFEU. In the alternative, if the applicant believes that EFSA committed maladministration in processing the confidentiality request, it has the right to lodge a complaint to the European Ombudsman pursuant to Article 228 TFEU.[116] It has been observed that it is the routine availability of companies' commercial information and data that presents a significant challenge in the application of Access Regulation.[117] As noted, having regard to Article 41 General Food Law, even if a document is granted confidential status under General Food Law or the relevant sectoral provisions, it may still be subject to requests for access to documents under Regulation (EC) No 1049/2001 on public access to documents held by Union institutions and Regulation (EC) No 1367/2006.

5.3 Some Reflections on Confidentiality and Innovation in the Aftermath of the Transparency Regulation

As a preliminary observation, it can be noted that the Practical Arrangements play a fundamental role in the implementation of the Transparency Regulation. The *Short selling* case has indeed established that agencies can be delegated powers if the underlying legislative framework is sufficiently detailed and their measures are subject to judicial review.[118] Moreover, it has been pointed out the necessity to coordinate agencies' legal framework with the adoption of post-legislative guidance by institutions.[119] There is no doubt that Practical Arrangements are expressions of

[115] Article 16 Practical Arrangements concerning transparency and confidentiality.

[116] EFSA (2021a), point 21, p. 47.

[117] Kocharov (2009), p. 341.

[118] See para. 5.2.

[119] Korkea-Aho and Leino (2017), p. 1071. About the Commission's discretion in adopting guidelines, see Judgment of the Court (Grand Chamber) of 19 July 2016, *Tadej Kotnik and Others v Državni zbor Republike Slovenije*, Case C-526/14, ECLI:EU:C:2016:570, para. 38 ff. On the expansion of the agency model and the increasing "delegation to autonomous structures, together

EFSA's discretion in the exercise of its powers. They are "binding means to interpret and implement the legal framework provided by the Transparency Regulation and by specifying the details for the implementation of required processes commit EFSA to the application of the Transparency Regulation".[120] Despite the clear indication of Practical Arrangements' legal basis in the text of the Transparency Regulation, scholars have criticised it for potential inconsistencies.[121] Such a complex framework may raise some doubts on EFSA's discretionary powers when it comes to significant constraints on private actors.

In relation to the *confidentiality requests* that an FBO can address to EFSA according to the new rules, one can note that the reaction period of two weeks within which a company can submit to EFSA the confirmatory application pursuant to Article 39b (2) General Food Law is potentially short in view of the "threatening consequences".[122] Once expired, EFSA publishes the part of the applications for which confidential treatment has not been acknowledged in administrative and legal proceedings.[123] Therefore, companies may face the consequences of the immediate publication of confidential data and the potential damages implied. Even in the case of *withdrawal* of the application for authorisation pursuant to Article 39d (3) General Food Law, the non-confidential part of the dossier submitted will already have been published. Therefore, competing companies already have the chance to gain access and take advantage of the information. The applicant must accept since the beginning the publication of certain information when deciding to carry out an application for authorising a regulated product to the market. Nonetheless, in the case of a decision in favour of the applicant, the Authority may *review* whether information that has been previously accepted as confidential may nevertheless be made public (Article 39c General Food Law). Consequently, businesses have to deal with potential disclosure of their trade secrets[124] not only when submitting the application, but at any time.[125] This may constitute an additional element of uncertainty for FBOs.

---

with a formalization of relationships and the creation of new layers of national and international regulation", see Leone (2021), p. 540.

[120] EFSA (2021a) Point 14, p. 42.

[121] Neusch et al. (2021), p. 198.

[122] Neusch et al. (2021), p. 196.

[123] Neusch et al. (2021), p. 197.

[124] The reference on trade secrets follows the definition by the Article 2 (1) Directive (EU) 2016/943 of the European Parliament and of the Council of 8 June 2016 on the protection of undisclosed know-how and business information (trade secrets) against their unlawful acquisition, use and disclosure, OJ L 157, 15.6.2016, p. 1–18: "'trade secret' means information which meets all of the following requirements: (a) it is secret in the sense that it is not, as a body or in the precise configuration and assembly of its components, generally known among or readily accessible to persons within the circles that normally deal with the kind of information in question; (b) it has commercial value because it is secret; (c) it has been subject to reasonable steps under the circumstances, by the person lawfully in control of the information, to keep it secret".

[125] Neusch et al. (2021), p. 197.

Businesses are subject to considerable constraints introduced by the Transparency Regulation, given that the confidentiality requests fall within the scope of the information items listed and in relation to a verifiable justification of the harm. While in the previous regime Recital 32 only provided that the information contained in the catalogue were confidential, now the burden of proof is reversed.[126] Only when the applicant proves imminent harm through disclosure the Authority would be allowed to take into account the confidentiality request. Indeed, the extent to which companies have to demonstrate *potential commercial harm* and the limited list provided in the legislation are of great discussion and concern among stakeholders.[127]

Moreover, when the new legal framework concerning marketing authorisation is read alongside the general rules on transparency and access under EU law, it becomes clear that the amended General Food Law is not perfectly aligned with the Access Regulation regime. The latter continues to apply to all documents held by EFSA. However, for confidentiality requests the General Food Law prescribes stringent criteria, while in the Access Regulation it is sufficient to prove the concrete and non-hypothetical harm to commercial interests. In the context of the publication of non-confidential environmental information included in application dossiers, the General Food Law now provides for even stricter requirements than the Aarhus framework.[128] In its quality of *lex specialis*, the General Food Law prevails over the Access Regulation and the Aarhus Regulation.

It is difficult to assess the impact of the disclosure of information on businesses. This regards not only the economic part that explains the *significance of the harm*, but also the *data* eligible for confidentiality. For instance, in the novel food framework, the detailed description of raw materials and production process is included in the list of data the applicant might ask to protect from publication. On the one hand, drafting a very specific description of ingredients might protect the FBO from the competition. On the other hand, the same piece of information may become subject to assessment, therefore information will be public.

Since EFSA shall launch a consultation on submitted studies (non-confidential version), stakeholders and the general public might indicate other relevant information on the substance at stake. Consequently, competitors may challenge a business's application also through the public consultation mechanism.

Among all the novelties of the General Food Law, the main concern by the side of the industry is that early disclosure of studies could let competitors gain an unfair market advantage over the initial applicant. As recalled before, Article 38 (1a) states that the public disclosure of applicants' supporting information should not be considered permission or license to use, reproduce, exploit the relevant data or information in breach of any IP rights. Moreover, it provides that the European Union shall not be responsible for its use by third parties. The Authority shall further ensure that "clear undertakings or signed statements are given to that effect by those who access

[126] Neusch et al. (2021), p. 195.

[127] Rusconi and Cozzolino (2021), p. 305. See de Boer et al. (2023), p. 327.

[128] See para. 5.2.1.

the relevant information prior to its disclosure". The breach of an IP right that may occur from the disclosure of the information or document that is claimed confidential "is not, in its own right, a sufficient justification for the award of confidential status".[129] Nonetheless, when the FBO provides a justification demonstrating the requirements of Article 10 of EFSA's Practical Arrangements on Transparency and Confidentiality, EFSA may award *confidential status* to the information with regard to which the request has been submitted. For instance, in the context of an authorisation request for a GM food a DNA sequence—that is also protected by IP rights—may get confidential status by means of complying with Article 10 Practical Arrangements concerning transparency and confidentiality.

Whereas FBOs consider patents as IP rights to protect their inventions, it is likely that in the coming years intellectual property will play a big role in food innovations. However, given the numerous constraints on FBOs introduced by the Transparency Regulation, it is still to be seen how this reform will affect the impact on innovation. For instance, in cases referred to the CJEU before the adoption of the Transparency Regulation, applicants sought interim measures to order EFSA[130] or EMA[131] to remove allegedly confidential information from websites.[132] Yet, it is probably too early to determine how the balance between transparency and confidentiality that the Regulation tries to achieve can be pursued.[133] However, as it has been pointed out, making confidential information public may expose FBOs to a competitive risk beyond the EU market. The applicant is protected in case of infringements of IP rights in the EU after proving them. Nonetheless, data protection and confidentiality do not extend beyond EU territory, and it remains unclear by which measures the applicants' rights are to be protected in the event of third parties' access where these rights are not bound by EU law.[134] It could be argued that within the new framework, administrative burdens on private actors may offer limited incentives for innovation in the EU. On this point, the Explanatory memorandum of the Transparency Regulation states that "it will not divert for innovation to defensive investments, since the measures are limited to providing transparent

[129] See EFSA (2021a) point 16, p. 43.

[130] See Order of the President of the General Court of 29 February 2016, *Chemtura Netherlands BV v European Food Safety Authority*, Case T-725/15 R, ECLI:EU:T:2016:128.

[131] Order of the President of the General Court of 25 April 2013, *AbbVie, Inc. and AbbVie Ltd v European Medicines Agency (EMA)*, Case T-44/13 R.

[132] Korkea-Aho and Leino (2017), p. 1077.

[133] The Annual Activity Report 2024 shows that EFSA issued 266 confidentiality decisions in 2023 and received only 1 confirmatory application revising the first decision (https://www.efsa.europa.eu/sites/default/files/2024-03/efsa-consolidated-annual-activity-report-2023_0.pdf).

[134] Korkea-Aho and Leino (2017), p. 1077; Neusch et al. (2021), p. 194; Jennings (2021), p. 175. According to Neusch et al., "it is not only the protection of the applicant's rights that is at issue, but also the intellectual property rights of third parties - for example, in the case of the publication of studies. The owner of the copyright to such a study is often a scientific publisher who has published it in one of his journals. Nevertheless, such studies must be submitted to the authority database as part of the application and published by the Authority".

information on studies that in any case the applicant has to carry out in accordance with already existing legislation".[135]

Notwithstanding the aim to redefine the risk assessment rules in a transparent way, it can be noted that the new rules on transparency were mostly framed by the EU legislator having in mind the "glyphosate saga" and the pesticide framework. In this area, the actors are often big companies, there is less competition overall, and the slow pace of innovation goes hand in hand with long term investments. On the contrary, in the food innovation space investments are rapid and involve medium businesses and start-ups acting in a highly competitive market.

For instance, it will be interesting to see how the new legislative framework will be enforced in the area of innovations such as cultured meat.[136] Cultured meat is animal meat produced with the help of tissue engineering techniques and is proposed as a solution to the "adverse effects of the meat producing industry" and its environmental impact.[137] In trying to make its way to the EU market, it is expected that cultured meat companies in the EU elect the novel food authorisation procedure when the production process applied allows for this route.[138]

Moreover, the increasing demand for investments in technology has an impact on the food sector. The development of new laws and regulations is required to create a "regulatory landscape that really fosters innovation".[139] This is of particular evidence in the area of new plant breeding techniques (NPBTs), where fast-moving technological development and uncertainties about their regulation create a "legal disruption".[140] When it comes to investments in food innovation, the value of a specific investment depends on the investments, the expected returns, and market risks

---

[135] See Proposal for a Regulation of the European Parliament and of the Council on the transparency and sustainability of the EU risk assessment in the food chain amending Regulation (EC) No 178/2002, Explanatory memorandum, p. 10. See also Schebesta et al. (2024), p. 76.

[136] On the broader topic concerning the interplay between innovation and regulatory framework, see Butenko and Larouche (2015), pp. 52–82.

[137] Ketelings et al. (2021), p. 1.

[138] Verzijden and Buijs (2020), p. 106. Ketelings et al. (2021), p. 5, report the industrial representatives' concerns about the authorisation procedure, especially regarding the possibility that EFSA asks additional information or the novelty of the product and its process which might introduce new food safety questions. In 2024, a French company has filed the first application to sell cultivated meat in Europe. See EFSA (2024).

[139] Purnhagen (2021), p. 7.

[140] Purnhagen (2021), p. 10. The debate surrounding NPBTs arise especially after the CJEU judgment which decided that plants developed by mutagenesis are regarded as GMOs (Judgment of the Court (Grand Chamber) of 25 July 2018, *Confédération paysanne and Others v Premier ministre and Ministre de l'agriculture, de l'agroalimentaire et de la forêt*, C-528/16, ECLI:EU:C:2018:583. On point, Garnett (2019), pp 349–355, Schebesta (2020). The European Commission has published a study regarding the status of New Genomic Techniques under Union law (Commission Staff Working Document "Study on the status of new genomic techniques under Union law and in light of the Court of Justice ruling in Case C-528/16", SWD (2021) 92 final, 29.4.2021). In July 2023, the European Commission proposed a new regulation on plants produced by certain new genomic techniques that allows an easier authorisation for those NGT plants that are considered equivalent to conventional plants (see Proposal for a New Regulation on plants produced by certain

related to the investment.[141] In this perspective, the new rules on transparency directly affect the phases of R&D and approval, most likely increasing their costs.

In the coming years, there might be room to verify not only whether *transparency* actually acts as a golden thread in the food sector, but also its potential link with the "innovation principle". Despite this principle is not based on primary or secondary legislation and does not exist in Member State legislation, it has gained increasing importance in EU policy.[142] In October 2013 the innovation principle was put for the first time into discussion by an industry association aiming to ensure that "whenever policy and regulatory decisions are under consideration the impact on innovation as a driver for jobs and growth should be assessed and addressed".[143] Industry associations claim that legislation focused on risk avoidance and elimination of scientific uncertainty may hamper technological innovations.[144] In this vein, companies advance the argument that "strict" regulatory frameworks such as the approval and authorisation scheme for pesticides with heavy scientific data requirements hinder innovation.[145] Consequently, they need higher investments in proving product safety for regulatory procedures rather than in innovation.

As specified on more occasions in the course of this book, in the market authorisation product information and transparency have a *public interests'* dimension. Under this perspective, the new rules are in line with the citizens' increasing demand for transparency and the recent judicial line on access to environmental information. The "legislative stream" from the Aarhus Convention and the recent decisions of the CJEU give priority to the right to demand access to safety-relevant information.[146] While it is not new for EU Courts the democratic and trust-enhancing dimension of transparency, it can be noted its "progressive expansion" towards information attributed to technical complexity.[147] Interestingly, relevant judgments of the CJEU are indeed taken into account by the EU legislator and EFSA in its Practical Arrangements. In its guidance documents, EFSA has reaffirmed that "established case law of the Court of Justice represents a binding authority providing the definitive interpretation of relevant Union law".[148]

Furthermore, applicants in confidentiality requests must either confirm that the information which is claimed confidential falls under the definition of

---

new genomic techniques COM (2023) 411 final). See also Fernández Albújar and Van der Meulen (2018), Karhrmann and Leggewie (2024).

[141] Wesseler and Purnhagen (2020), p. 7.

[142] Garnett et al. (2018), p. 1.

[143] Garnett et al. (2018), p. 2. See European Risk Forum (2013); European Risk Forum (2015). The European Commission in November 2019 delivered a study on the evaluation of the innovation principle, see European Commission et al. (2019).

[144] For instance, BusinessEurope, the European Risk Forum and the European Round Table of Industrialists 2015.

[145] Röttger-Wirtz (2018), p. 2. See European Risk Forum (2020), p. 34.

[146] Autio et al. (2023), p. 189.

[147] Morvillo (2019), p. 426.

[148] EFSA (2021a) point 15, p. 43.

"environmental information" or explain why the information does not fall under such definition.[149]

These elements argue in favour of a *unitary view on transparency* that connects EU food law as a sectoral law stemming from EU law. As described, in the General Food Law the principle of transparency is expressly enshrined in Articles 9 and 10, while the transparency of the Authority's action is established in Article 38. However, several transparency obligations nonetheless arise in *other provisions*, especially in conjunction with the *businesses' responsibilities*. As observed in the GFL's analysis, the "transparency dimensions" described appear in a considerable number of legal provisions.[150] Consequently, the various interests implied must be connected in an effective manner. As seen in this chapter, the implementation of the Transparency Regulation must take into account the balance between transparency on the one hand, and the interests involved on the other hand. Both commercial confidentiality on the side of FBOs and the stakeholders' interests in innovation in the food sector are challenges to be faced in these years.

References

Ankersmit L (2018) Case C-57/16P ClientEarth v Commission: Citizen's Participation in EU Decision-Making and the Commission's Right of Initiative. European Law Blog. Available via https://www.europeanlawblog.eu/pub/case-c-57-commissions-right-of-initiative/release/1

Arcuri A, Hale Hendlin YH (2019) The chemical anthropocene: glyphosate as a case study of pesticide exposures. King's Law J 30(2):234–253

Autio S, Koëter H, Carretero M et al (2023) Assessing assessors: proposal for a guidance for evaluating the scientific performance of a pesticide regulatory authority. Eur J Risk Regul 14(1):178–190

Buonsante VA, Friel F (2017) What is information relating to emissions into the environment? Eur J Risk Regul 8:453–460

BusinessEurope, the European Risk Forum and the European Round Table of Industrialists (2015) Better Framework for Innovation. Fuelling EU policies with an Innovation Principle. Available at https://www.eriforum.eu/uploads/2/5/7/1/25710097/businesseurope-erf-ert_innovation_principle_joint_statement.pdf. Accessed 18 Apr 2025

Butenko A, Larouche P (2015) Regulation for innovativeness or regulation of innovation. Law Innov Technol 7:52–82

Chatzopoulou S, Eriksson NL, Eriksson D (2020) Improving Risk Assessment in the European Food Safety Authority: lessons from the European Medicines Agency. Front Plant Sci 6(11):349

Client Earth (2018) Reform of the General Food Law. Analysis of the new provisions on transparency, August 2018. Available at: https://www.clientearth.org/latest/documents/reform-of-the-general-food-law-analysis-of-the-new-provisions-on-transparency/. Accessed 18 Apr 2025

Curtin D, Leino-Sandberg P (2016) Openness, transparency and the right of access to documents in the EU. An in-depth analysis. Robert Schuman Centre for Advanced Studies Research Paper No. RSCAS 2016/63: 1-27. Available via https://ssrn.com/abstract=2879315

de Boer A, Morvillo M, Röttger-Wirtz S (2023) Fragmented transparency: the visibility of agency science in European Union risk regulation. Eur J Risk Regul 14:313–331

[149] Article 10 Practical Arrangements concerning transparency and confidentiality.

[150] See Chap. 4, para. 4.5.

Donati A (2020), Arrêt Antony C. Tweedale/ Autorité européenne de sécurité des aliments: la transparance de la procédure d'evaluation du glyphosate à l'épreuve du Tribunal de l'Union européenne. Revue européenne de droit de la consommation (R.E.D.C.) 2:309–316
EFSA (2014) Discussion Paper - Transformation to an "Open EFSA", 17 July 2014. Available at https://www.efsa.europa.eu/en/corporate/pub/openefsadiscussionpaper14#documents. Accessed 18 Apr 2025
EFSA (2021a) Questions and Answers on EFSA Practical Arrangements. Available at https://www.efsa.europa.eu/sites/default/files/2021-03/questions-and-answers-efsa-practical-arrangements.pdf. Accessed 18 Apr 2025
EFSA (2021b) Administrative guidance for the processing of applications for regulated products (update 2021), 3 March 2021, Available at https://www.efsa.europa.eu/en/supporting/pub/en-6471. Accessed 18 April 2025
EFSA (2021c) Administrative guidance for the preparation of applications on genetically modified plants. EFSA Supporting publications 18:3 https://doi.org/10.2903/sp.efsa.2021.EN-6474
EFSA et al (2024) EFSA Scientific Colloquium: Cell Culture-derived Foods and Food Ingredients. Available via https://efsa.onlinelibrary.wiley.com/doi/10.2903/sp.efsa.2024.EN-8664
European Commission: Directorate-General for Research and Innovation, CEPS, Simonelli F, Renda A (2019) Study supporting the interim evaluation of the innovation principle – Final report. Available via https://data.europa.eu/doi/10.2777/620609. Accessed 18 Apr 2025
European Risk Forum (2013) The Innovation Principle. Letter to the Presidents of the European Commission, the European Council and the European Parliament. Available at https://corporateeurope.org/sites/default/files/corporation_letter_on_innovation_principle.pdf. Accessed 18 Apr 2025
European Risk Forum (2015) European Risk Forum Monograph. Fostering Innovation Better Management of Risk. Available at https://www.eriforum.eu/uploads/2/5/7/1/25710097/monograph_innovation_principle.pdf. Accessed 18 Apr 2025
Fernández Albújar G, van der Meulen B (2018) The EU's GMO Concept: Analysis of the GMO Definition in EU law in the Light of New Breeding Techniques (NBTs). Eur Food Feed Law Rev 13(1):14–28
Gabbi S, Wood M, Strauss B (2020) Controlling the European Food Safety Authority. In: Sholten M, Brenninkmeijer A (eds) Controlling EU agencies. Edward Elgar Publishing, Cheltenham and Camberley, pp 196–214
Garnett K (2019) Hold your pipettes: The European Court of Justice's findings in Confédération Paysanne & Others stirs GMOtions. Rev Eur Comp Int Environ Law 28(3):349–355
Garnett K, Van Calster G, Reins L (2018) Towards an innovation principle: an industry trump or shortening the odds on environmental protection? Law Innov Technol 10(1):1–14
Giulimondi F (2015) Corte di Giustizia dell'Unione europea (Sezione seconda), sentenza 16 luglio 2015, causa C-615/13 P, ClientEarth e Pesticide Action Network Europe (PAN Europe) c. European Food Safety Authority (EFSA). Diritto di accesso e trasparenza: veri baluardi della credibilità delle Istituzioni europee. DPCE Online 24(4):171–180
Hickey E, Weimer M (2022) The transparency of EU agency science – towards a new proactive approach. Common Mark Law Rev 59(3):673–710
Holle M (2014) The protection of proprietary data in novel foods – how to make it work. Eur Food Feed Law Rev 9(5):280–284
Jenning S (2021) Transparency in European Union food law. In: Steier G, Friedlander A (eds) Food system transparency: law, science and policy of food and agriculture. Advances in agroecology. CRC Press, Boca Raton, pp 161–179
Karhrmann J, Leggewie G (2024) European Commission's plans for a special regulation of plants created by new genomic techniques. Eur Pap 9(1):21–38
Ketelings L, Kremers S, de Boer A (2021) The barriers and drivers of a safe market introduction of cultured meat: a qualitative study. Food Control 130:108299. https://doi.org/10.1016/j.foodcont.2021.108299

Kocharov A (2009) Data ownership and access rights in the European food safety authority. Eur Food Feed Law Rev 4(5):335–346

Korkea-Aho E, Leino P (2017) Who owns the information held by EU agencies: weed killers, commercially sensitive information and transparent and participatory governance. Common Mark Law Rev 54(4):1059–1092

Leone L (2021) EFSA under revision: transparency and sustainability in the food chain. Yearb Eur Law 39:536–568

Mendes J (2011) Participation in EU rule-making: a rights-based approach. Oxford University Press, Oxford

Mendes J (2020) The principle of transparency and access to documents in the EU: for what, for whom and of what? Universitè Du Luxembourg, Faculty of Law, Economics and Finance Law Working Paper Series, no. Paper number 2020-004. Available via https://www.ssrn.com/abstract=3557795

Monaco A (2023) Data protection under the novel food regulation: valuable instrument or barrier to innovation? Insights from the insect sector. Eur Food Feed Law Rev 18(3):172–178

Morvillo M (2019) The general court orders disclosure of glyphosate-related scientific studies: Tweedale, Hautala, and the concept of environmental information in the context of plant protection products. Eur J Risk Regul 10:419–427

Neusch A, Orth A-M, Meisterernst A (2021) New rules for applications to EFSA - changes due to the transparency regulation. Eur Food Feed Law Rev 16(3):188–201

Purnhagen K (2021) You want it extra Crispery?: Legal disruption through new plant breeding technologies in the EU. Yearb Eur Law 40:374–397

Röttger-Wirtz S (2020) D.2.3 Glyphosate case study: Contribution to the REconciling sCience, Innovation and Precaution through the Engagement of Stakeholders Project (RECIPES). Horizon 2020 Framework Programme of the European Union under Grant Agreement no. 824665. Available via Maastricht University. https://recipes-project.eu/sites/default/files/2020-10/_CS6_Glyphosate.pdf

Rusconi G, Cozzolino G (2021) Transparency regulation and implementation issues. Eur Food Feed Law Rev 16(4):303–308

Sapienza S (2019). Transparency and openness in food safety: insights about new data confidentiality rules. Available via https://doi.org/10.5281/zenodo.2766359

Schebesta H (2020) Confédération paysanne case (C-528/16): Legal Perspective on the GMO Judgment of the European Court of Justice. Revue européenne de droit de la consommation (R.E.D.C.) 2:369–378

Schebesta H, Alessandrini M, Cazzini F, Rolandi S, Hillesheim A, Vos E, Helmlinger M, Calo A, Mezzacapo E, Delhomme V, Fortin N, Milian Gomez J, Macchi C, Tracz M, Coli F, Paganizza V, Vanuzzo M, de Jager C, Pierson-Brown T (2024) The future of food law: synopsis article and research agenda. Rivista di diritto alimentare 18(1):62–81

Simpson C (2016) Data protection under food law post the novel foods regulation. Eur Food Feed Law Rev 11(4):309–314

Van Maldegem K (2016) EU Court of Justice landmark rulings on access to environmental information. Available via https://www.fieldfisher.com/en/insights/eu-court-of-justice-landmark-rulings-on-access-to-environmental-information-1

Verzijden K, Buijs J (2020) Meat 3.0 — How Cultured Meat is Making its Way to the Market. Eur Food Feed Law Rev 15(2):96–107

Volpato A (2019) Judicial review of the acts of EU agencies: discretion escaping scrutiny? CERiM Paper Series: CERiM Online Paper 1/2019, Available via https://doi.org/10.2139/ssrn.3500731. Accessed 18 Apr 2025

Vos E (2005) Independence, accountability and transparency of european regulatory agencies. In: Gerardin D, Muñoz R, Petit N (eds) Regulation through agencies in the EU. A new paradigm of European governance. Edward Elgar Publishing, Cheltenham and Camberley, pp 120–140

Vos E, Volpato A, Bellenghi G (2023) Independence and transparency policies of the European Food Safety Authority (EFSA), publication for the Committee on Environment, Public Health and

Food Safety (ENVI), Policy Department for Economic, Scientific and Quality of Life Policies, European Parliament, Luxembourg. Available at https://www.europarl.europa.eu/RegData/etudes/STUD/2023/740080/IPOL_STU(2023)740080_EN.pdf. Accessed 7 May 2025

Wesseler JHH, Purnhagen KP (2020) EU Regulation of New Plant Breeding Technologies and Their Possible Economic Implications for the EU and Beyond. Campus Kulmbach Legal Working Papers 1/20

Westergrent A (2016) The data liberation movement: regulation of clinical sharing in the European Union and the United States. Houston J Int Law 38:887–910

World Intellectual Property Organization (WIPO) (2018) A Guide to Intellectual Property Issues in Access and Benefit-sharing Agreements. Available at https://www.wipo.int/edocs/pubdocs/en/wipo_pub_1052.pdf. Accessed 18 Apr 2025

Chapter 6
Conclusions

Abstract This chapter presents the main conclusions and final reflections.

This concluding chapter offers the following reflections. As noted at the beginning of the book, transparency has emerged as the norm or legal principle connecting the changing social realities with the law.[1] Being a broad concept, transparency in law encompasses several "dimensions" or "faces". Transparency enhances democracy, legitimacy, quality of governance and accountability, as well as improves the realisation of individual rights and market efficiency.[2] The framework described in Chap. 2 shows how each dimension of transparency serves different purposes, facilitating a deeper understanding of the complex topic.

Moreover, adopting a broader view on transparency and offering a conceptual analysis of its dimensions has made clear that transparency obligations do not only address public authorities but also private entities. This is true particularly in the food sector, where FBOs are pivotal actors in the food chain.

In the attempt to define transparency, Chap. 2 also provides a definition that focuses on the "instrumental" value of transparency. In this book, transparency is understood as the *availability and exchangeability of information* between different actors in a specific context[3]. The adaptability of the notion to multiple contexts is valuable when investigating the existence of a general principle of transparency in European Union law—and food law specifically.

As concluded in Chap. 3, under the European Union law framework transparency has been mainly addressed in the vertical relationship between public authorities and citizens. Specifically, the doctrine has distinguished between *active models*, in which institutions take initiative in releasing information to the public, and *passive models,* where the information is directly requested by the public. Especially after the Lisbon Treaty, transparency has featured as a *general objective of the EU*

[1] See Chap. 2.

[2] See Chap. 2.

[3] See Chap. 2.

M. Vanuzzo, *Transparency in EU Law*, Studies in European Economic Law and Regulation 29, https://doi.org/10.1007/978-3-032-02345-2_6

and has been connected with dialogue and consultation with civil society. Therefore, when it comes to the legislative framework encompassing transparency in EU law, it becomes clear that the discussion around transparency issues mostly covers obligations imposed on public bodies.[4]

In this vein, citizens aiming to participate in the democratic life of the European Union invoke *access to documents* as a pivotal legal tool.

Despite recognising transparency as an important legal principle, scholars still confirm that in the EU law landscape, *only access to documents* can be viewed as a *general principle of EU law*.[5] As highlighted in Chap. 3, transparency cannot be declared a general principle of EU law unless is intended in the sense of access to documents.

As far as food law is concerned, the target of transparency obligations is private companies qualified as FBOs, together with EFSA acting as a public authority. In this sector, public debate on the lack of transparency after the BSE crisis and the subsequent debate led to the publication of the 2000 White Paper on Food Safety, followed by the adoption of the General Food Law[6]. The General Food Law has been amended by the Transparency Regulation, which aimed to address the shortcomings of the General Food Law by introducing several provisions on transparency. In Chap. 4, the analysis attempted to potentially discover a principle of transparency that encompasses the General Food Law by applying the conceptual dimensions of transparency outlined in Chap. 2. Interestingly, it was found that all the "transparency dimensions" can be uncovered by linking them to the transparency obligations. This occurrence might lead to the development of an overarching principle of transparency in EU food law. The "transparency dimensions" are present not only in the new provisions introduced by the Transparency Regulation, but also in several articles of the General Food Law that were not amended by the reform.

The analysis of transparency in the food sector has been conducted by separating the doctrinal view from the CJEU approach. Delving into the case law related to transparency, Chap. 4 reveals that the Court primarily refers to transparency mostly in the context of the *risk assessment principle*. As noted in the General Food Law analysis, risk assessment is at the core of the food safety regulation, and it appears that the Court has relied on it whenever *transparent* decision-making was required. This result is confirmed by the fact that the Transparency Regulation was adopted with the specific aim of *increasing transparency in the risk assessment*. As seen, the purpose of the new rules is to ensure transparency in all studies and information provided by the industry during the risk assessment process.

From a food sector standpoint, the aim of safeguarding consumer interests, ensuring food safety, and promoting trade highlights the pivotal role that transparency can play. In this light and beyond, it may operate as a "golden thread" through EU food law, in accordance with the White Paper on Food Safety. Nevertheless, as

[4] See Chap. 3.

[5] See Chap. 3.

[6] See Chap. 4.

concluded in Chap. 4, the CJEU lacks a consistent approach to what transparency entails in this context. Currently, there is a limited understanding of how it operates and several functions of transparency are not being fulfilled. The judicial reasoning in the analysed case law contributes little to the development of an overarching principle.

The novelties introduced by the Transparency Regulation might be undermined if the interpretation of legal provisions lacks connection and reasoning in terms of principles and general rules.

Finally, in both EU law in general and EU food law specifically, the CJEU has connected the principle of transparency to other well-established principles of law. In this vein, as deduced in Chap. 3 regarding EU law in general and Chap. 4 in the case law analysis of the food sector, it appears that transparency does not function as a self-standing principle, but rather relies on other principles.[7]

For the first time in the history of the General Food Law, transparency is strengthened despite the absence of an evident food safety risk at *an early stage of the risk assessment*. However, the new regulatory landscape has imposed several constraints on private actors, as identified in Chap. 5.[8] In exploring how the principle of transparency can be applied in or to the *private sector*, it was found that in the aftermath of the Transparency Regulation the dividing line between transparency and confidentiality necessary to protect FBOs' commercial interests remains blurred. Moreover, in Chap. 5 the analysis shows that when the new legal framework is connected with the general rules on transparency and access in EU law, the General Food Law does not fully align with the Access Regulation regime. The latter continues to apply to all documents held by EFSA. Nevertheless, the Transparency Regulation imposes stricter rules for granting confidentiality, which might constitute a burden on FBOs. In the Access Regulation, it is sufficient to prove the concrete and non-hypothetical harm to commercial interests, instead.[9] Although transparency is often regarded as a means to strengthen consumers' confidence, it remains to be determined whether the reform effectively enhances their trust.

In addressing transparency, this book adopted a legal approach centred on risk assessment, drawing on the novelties introduced by the reform of the General Food Law. However, other interesting research avenues on transparency in food law might consider a market-based approach focusing on EU internal market and the broader WTO context. In EU food law, this might lead to exploring the Regulation on Official Controls[10] and its enforcement to assess whether transparency is a lever or

[7] See also Chaps. 3 and 4.

[8] See Chap. 5.

[9] See Chap. 5.

[10] Regulation (EU) 2017/625 of the European Parliament and of the Council of 15 March 2017 on official controls and other official activities performed to ensure the application of food and feed law, rules on animal health and welfare, plant health and plant protection products, amending Regulations (EC) No 999/2001, (EC) No 396/2005, (EC) No 1069/2009, (EC) No 1107/2009, (EU) No 1151/2012, (EU) No 652/2014, (EU) 2016/429 and (EU) 2016/2031 of the European Parliament and of the Council, Council Regulations (EC) No 1/2005 and (EC) No 1099/2009 and

a limitation for the market. Furthermore, transparency in food law might be discussed using a supply-chain based approach. This perspective is arguably linked with transparency in the meaning of "information to consumers".[11] While it can be expected that such an approach relies on policy instruments, EU consumer law and the Regulation on the provision of food information to consumers (FIC) can constitute a solid legal foundation.

Expanding the understanding of transparency through principle-oriented reasoning and providing clearer definitions not only promotes a greater interest in the EU among citizens but also ensures more certainty for businesses operating in the food sector. The confidence of FBOs in regulatory procedures and legal tools serves as a significant lever for the functioning of the EU market. In the future, transparency could play a role in shaping more flexible legislation and fostering principle-based reasoning in interpreting food law, balancing public and private interests effectively.

Council Directives 98/58/EC, 1999/74/EC, 2007/43/EC, 2008/119/EC and 2008/120/EC, and repealing Regulations (EC) No 854/2004 and (EC) No 882/2004 of the European Parliament and of the Council, Council Directives 89/608/EEC, 89/662/EEC, 90/425/EEC, 91/496/EEC, 96/23/EC, 96/93/EC and 97/78/EC and Council Decision 92/438/EEC (Official Controls Regulation) Text with EEA relevance. OJ L 95, 7.4.2017, pp. 1–142.

[11] Regulation (EU) No 1169/2011 of the European Parliament and of the Council of 25 October 2011 on the provision of food information to consumers, amending Regulations (EC) No 1924/2006 and (EC) No 1925/2006 of the European Parliament and of the Council, and repealing Commission Directive 87/250/EEC, Council Directive 90/496/EEC, Commission Directive 1999/10/EC, Directive 2000/13/EC of the European Parliament and of the Council, Commission Directives 2002/67/EC and 2008/5/EC and Commission Regulation (EC) No 608/2004, OJ L 304, 22.11.2011, pp. 18–63.

Table of Cases

Court of Justice of the European Union (CJEU)

Court of Justice

Judgment of the Court of 13 June 1958, *Meroni & Co., Industrie Metallurgiche, SpA v High Authority of the European Coal and Steel Community*, C-9/56, ECLI:EU:C:1958:7.

Judgment of the Court of 20 February 1979, *Rewe-Zentral AG v Bundesmonopolverwaltung für Branntweinudgment*, C-120/78, ECLI:EU:C:1979:42.

Judgment of the Court (First Chamber) of 14 May 1981, *Giuseppe Romano v Institut national d'assurance maladie-invalidité*, C-98/80, ECLI:EU:C:1981:104.

Judgment of the Court of 9 December 1981, *Commission of the European Communities v Italian Republic*, C-193/80, ECLI:EU:C:1981:298.

Judgment of the Court (First Chamber) of 15 March 1984, *Tradax Graanhandel BV v Commission of the European Communities*, C-64/82, ECLI:EU:C:1984:106.

Judgment of the Court (Fourth Chamber) of 10 July 1985, *Wirtschaftsvereinigung Eisen-und Stahlindustrie v Commission of the European Communities*, Case 27/84, ECLI:EU:C:1985:311.

Judgment of the Court of 23 April 1986, *Parti écologiste "Les Verts" v European Parliament*, Case 294/83, ECLI:EU:C:1986:166.

Judgment of the Court (Fifth Chamber) of 23 September 1986, *AKZO Chemie BV and AKZO Chemie UK Ltd v Commission of the European Communities*, C-5/85, ECLI:EU:C:1986:328.

Judgment of the Court of 28 November 1991, *Bureau Européen des Unions de Consommateurs (BEUC) v Commission of the European Communities*, C-170/89, ECLI:EU:C:1991:450.

M. Vanuzzo, *Transparency in EU Law*, Studies in European Economic Law and Regulation 29, https://doi.org/10.1007/978-3-032-02345-2

Judgment of the Court (Fifth Chamber) of 25 April 1996, *Commission of the European Communities v Kingdom of Belgium*, Case C-87/94, ECLI:EU:C:1996:161.

Judgment of the Court of 30 April 1996, *Kingdom of the Netherlands v Council of the European Union*, C-58/94, ECLI:EU:C:1996:171.

Judgment of the Court of 5 May 1998, *The Queen v. Ministry of Agric., Fisheries and Food,* C-157/96, ECLI:EU:C:1998:191.

Judgment of the Court of 5 May 1998,, *U.K. Gr. Brit and N. Ir. V. Commission,* C-180/96, ECLI:EU:C:1998:192.

Judgment of the Court (Fifth Chamber) of 1 October 1998, *U.K. Gr. Brit and N. Ir. V. Commission*, C-209/96, ECLI:EU:C:1998:448.

Judgment of the Court (First Chamber) of 22 April 1999, *Kernkraftwerke Lippe-Ems GmbH v Commission of the European Communities*, C-161/97, ECLI:EU:C:1999:193.

Judgment of the Court (First Chamber) of 18 November 1999, *Unitron Scandinavia A/S and 3-S A/S, Danske Svineproducenters Serviceselskab v Ministeriet for Fødevarer, Landbrug og Fiskeri*, C-275/98, ECLI:EU:C:1999:567.

Judgment of the Court of 23 November 1999, *Portuguese Republic v Council of the European Union*, C-149/96, ECLI:EU:C:1999:574.

Judgment of the Court (Sixth Chamber) of 7 December 2000, *Telaustria Verlags GmbH and Telefonadress GmbH v Telekom Austria AG*, Case C-324/98, ECLI:EU:C:2000:669.

Judgment of the Court (Fifth Chamber) of 13 September 2001, *Commission of the European Communities v Kingdom of Spain*, C-417/99, ECLI:EU:C:2001:445.

Judgment of the Court of 13 December 2001, *Commission of the European Communities v French Republic*, C-1/00, ECLI:EU:C:2001:687.

Judgment of the Court (Sixth Chamber) of 7 March 2002, *Italian Republic v Commission of the European Communities*, C-310/99, ECLI:EU:C:2002:143.

Judgment of the Court (Sixth Chamber) of 18 June 2002, *Kingdom of Spain v Commission of the European Communities*, C-398/00, ECLI:EU:C:2002:382.

Judgment of the Court (Sixth Chamber) of 12 December 2002, *Universale-Bau AG, Bietergemeinschaft: 1) Hinteregger & Söhne Bauges.m.b.H. Salzburg, 2) ÖSTÜ-STETTIN Hoch- und Tiefbau GmbH v Entsorgungsbetriebe Simmering GmbH*, C-470/99, ECLI:EU:C:2002:746.

Judgment of the Court of 20 May 2003, *Ravil SARL v Bellon import SARL and Biraghi SpA*, C-469/00, ECLI:EU:C:2003:295.

Judgment of the Court of 20 May 2003, *Consorzio del Prosciutto di Parma and Salumificio S. Rita SpA v Asda Stores Ltd and Hygrade Foods Ltd.*, Case C-108/01, ECLI:EU:C:2003:296.

Judgment of the Court (Sixth Chamber) of 29 April 2004, *Commission of the European Communities v CAS Succhi di Frutta SpA*, Case C-496/99, ECLI:EU:C:2004:236.

Judgment of the Court (First Chamber) of 11 January 2005, *Stadt Halle and RPL Recyclingpark Lochau GmbH v Arbeitsgemeinschaft Thermische Restabfall- und Energieverwertungsanlage TREA Leuna,* C-26/03, ECLI:EU:C:2005:5.

Judgment of the Court (Third Chamber) of 14 April 2005, *Kingdom of Belgium v Commission of the European Communities*, C-110/03, ECLI:EU:C:2005:223.

Judgment of the Court (Grand Chamber) of 12 July 2005, *The Queen, on the application of Alliance for Natural Health and Nutri-Link Ltd v Secretary of State for Health* (C-154/04) and *The Queen, on the application of National Association of Health Stores and Health Food Manufacturers Ltd v Secretary of State for Health and National Assembly for Wales* (C-155/04), Joined cases C-154/04 and C-155/04, ECLI:EU:C:2005:449.

Judgment of the Court (Grand Chamber) of 21 July 2005, *Consorzio Aziende Metano (Coname) v Comune di Cingia de' Botti*, C-231/03, ECLI:EU:C:2005:487.

Judgment of the Court (Grand Chamber) of 6 December 2005, *The Queen, on the application of ABNA Ltd and Others v Secretary of State for Health and Food Standards Agency* (C-453/03), *Fratelli Martini & C. SpA and Cargill Srl v Ministero delle Politiche Agricole e Forestali and Others* (C-11/04), *Ferrari Mangimi Srl and Associazione nazionale tra i produttori di alimenti zootecnici (Assalzoo) v Ministero delle Politiche Agricole e Forestali and Others* (C-12/04) and *Nederlandse Vereniging Diervoederindustrie (Nevedi) v Productschap Diervoeder* (C-194/04), Joined cases C-453/03, C-11/04, C-12/04 and C-194/04, ECLI:EU:C:2005:741.

Judgment of the Court (First Chamber) of 13 October 2005, *Parking Brixen GmbH v Gemeinde Brixen and Stadtwerke Brixen* AG, C-458/03, ECLI:EU:C:2005:605.

Judgment of 6 March 2006, *Interporc v Commission*, C-41/00P, ECLI:EU:C:2003:125.

Judgment of the Court (Second Chamber) of 23 November 2006, *Lidl Italia Srl v Comune di Arcole (VR),* Case C315/05, ECLI:EU:C:2006:736.

Judgment of the Court (Grand Chamber) of 18 December 2007, *Kingdom of Sweden v Commission of the European Communities and Others*, Case C-64/05 P, ECLI:EU:C:2007:802.

Judgment of the Court (Grand Chamber) of 1 July 2008, *Kingdom of Sweden and Maurizio Turco v Council of the European Union*, Joined cases C-39/05 P and C-52/05 P, ECLI:EU:C:2008:374.

Judgment of the Court (Third Chamber), 21 June 2012, *IFAW Internationaler Tierschutz-Fonds gGmbH v European Commission*, Case C-135/11 P, ECLI:EU:C:2012:376.

Judgment of the Court (Fourth Chamber), 11 April 2013, *Karl Berger v Freistaat Bayern*, C-636/11, ECLI:EU:C:2013:227.

Judgment of the Court (Grand Chamber), 22 January 2014, *United Kingdom of Great Britain and Northern Ireland v European Parliament and Council of the European Union*, C-270/12, ECLI:EU:C:2014:18.

Judgment of the Court (Second Chamber) of 16 July 2015, *ClientEarth and Pesticide Action Network Europe (PAN Europe) v European Food Safety Authority*, Case C-615/13, ECLI:EU:C:2015:489.

Judgment of the Court (Fourth Chamber) of 17 December 2015, *Société Neptune Distribution v Ministre de l'Économie et des Finances*, C-157/14, ECLI:EU:C:2015:823.

Judgment of the Court (Fifth Chamber) of 23 November 2016, *European Commission v Stichting Greenpeace Nederland and Pesticide Action Network Europe (PAN Europe)*, C-673/13 P, ECLI:EU:C:2016:889.

Judgment of the Court (Fifth Chamber) of 23 November 2016, *Bayer CropScience SA-NV and Stichting De Bijenstichting v College voor de toelating van gewasbeschermingsmiddelen en biociden*, C-442/14, ECLI:EU:C:2016:890.

Judgment of the Court (Grand Chamber) of 19 July 2016, Tadej Kotnik and Others v Državni zbor Republike Slovenije, Case C-526/14, ECLI:EU:C:2016:570.

Judgment of the Court (Fourth Chamber) of 19 January 2017, *Queisser Pharma GmbH & Co. KG v Bundesrepublik Deutschland*, C-282/15, ECLI:EU:C:2017:26.

Judgment of the Court (Grand Chamber) of 25 July 2018, *Confédération paysanne and Others v Premier ministre and Ministre de l'agriculture, de l'agroalimentaire et de la forêt*, C-528/16, ECLI:EU:C:2018:583.

Judgment of the Court (Grand Chamber) of 4 September 2018, *ClientEarth v European Commission*, Case C-57/16 P, ECLI:EU:C:2018:660.

Judgment of the Court (Fourth Chamber) of 22 January 2020, *PTC Therapeutics International Ltd v European Medicines Agency*, C-175/18 P, ECLI:EU:C:2020:23.

Judgment of the Court (Second Chamber) of 4 October 2024, *Protéines France and Others v Ministre de l'Économie, des Finances et de la Souveraineté industrielle et numérique*, C-438/23, ECLI:EU:C:2024:826.

Court of First Instance/General Court

Judgment of the Court of First Instance (Second Chamber, extended composition) of 19 October 1995, *John Carvel and Guardian Newspapers Ltd v Council of the European Union*, Case T-194/94, ECLI:EU:T:1995:183.

Judgment of the Court of First Instance (Second Chamber, extended composition) of 22 May 1996, *Associazione Italiana Tecnico Economica del Cemento (AITEC) v Commission of the European Communities*, Case T-277/94, ECLI:EU:T:1996:66.

Judgment of the Court of First Instance (First Chamber, extended composition) of 18 September 1996, *Postbank NV v Commission of the European Communities*, Case T-353/94, ECLI:EU:T:1996:119.

Judgment of the Court of First Instance of 5 March 1997, *WWF UK (World Wide Fund for Nature) v Commission of the European Communities*, T-105/95, ECLI:EU:T:1997:26.

Judgment of the Court of First Instance (Fourth Chamber, extended composition) of 17 June 1999, *Associação dos Refinadores de Açúcar Portugueses (ARAP), Alcântara Refinarias - Açúcares SA and RAR Refinarias de Açúcar Reunidas SA v Commission of the European Communities*. Case T-82/96, ECLI:EU:T:1999:127.

Judgment of the Court of First Instance (First Chamber) of 19 July 1999, *Heidi Hautala v Council of the European Union*, T-14/98, ECLI:EU:T:1999:157.

Court of First instance in Judgment of 14 October 1999 *CAS Succhi di Frutta SpA v Commission of the European Communities* Joined cases T-191/96 and T-106/97, ECLI:EU:T:1999:256.

Judgement of the Court of First Instance (Third Chamber) 11 September 2002, *Pfizer Animal Health SA v Council of the European Union*, T-13/99, ECLI:EU:T:2002:209.

Judgment of the Court of First Instance (Third Chamber) of 11 September 2002, *Alpharma Inc. v Council of the European Union*, T-70/99, ECLI:EU:T:2002:210.

Judgment of the Court of First Instance (Second Chamber, extended composition) of 26 November 2002, *Artegodan GmbH and Others v Commission of the European Communities,* Joined cases T-74/00, T-76/00, T-83/00, T-84/00, T-85/00, T-132/00, T-137/00 and T-141/00, ECLI:EU:T:2002:283.

Judgment of the Court of First Instance (Second Chamber) of 10 March 2004, *Malagutti-Vezinhet SA v Commission of the European Communities*, Case T-177/02, ECLI:EU:T:2004:72.

Judgment of the Court (Third Chamber) of 1 April 2004, *Bellio F.lli Srl v Prefettura di Treviso*, C-286/02, ECLI:EU:C:2004:212.

Judgment of the Court of First Instance (Fifth Chamber, extended composition) of 30 November 2004, *IFAW Internationaler Tierschutz-Fonds gGmbH v Commission of the European Communities*, Case T-168/02, ECLI:EU:T:2004:346.

Judgment of the Court of First Instance (First Chamber, extended composition) of 13 April 2005, *Verein für Konsumenteninformation v Commission of the European Communities*, Case T-2/03, ECLI:EU:T:2005:125.

Judgment of the Court of First Instance (First Chamber) of 13 December 2006, *Julia Abad Pérez and Others v Council of the European Union and Commission of the European Communities*, T-304/01, ECLI:EU:T:2006:389.

Judgment of the Court of First Instance (First Chamber) of 13 December 2006; *É. R., O. O., J. R., A. R., B. P. R. and Others v Council of the European Union and Commission of the European Communities*, T-138/03, ECLI:EU:T:2006:390.

Judgment of the Court of First Instance (Third Chamber) of 30 January 2008, *Terezakis v Commission of the European Communities*, T-380/04, ECLI:EU:T:2008:19.

Judgment of the General Court (Eighth Chamber) of 12 June 2015, *The Health Food Manufacturers' Association and Others v European Commission*, T-296/12, ECLI:EU:T:2015:375.

Judgment of the General Court (Fifth Chamber) of 16 March 2016, *Dextro Energy GmbH & Co. KG v European Commission*, T-100/15, ECLI:EU:T:2016:150.

Judgment of the General Court (Third Chamber) of 20 September 2016, *Pesticide Action Network Europe (PAN Europe) v European Commission*, Case T-51/15, ECLI:EU:T:2016:519.

Judgment of the General Court (Seventh Chamber, Extended Composition) of 22 March 2018, *Emilio De Capitani v European Parliament*, Case T-540/15, ECLI:EU:T:2018:167

Judgment of the General Court (First Chamber, Extended Composition) of 17 May 2018, *BASF Agro BV and Others v European Commission*, T-584/13, ECLI:EU:T:2018:279.

Judgment of the General Court (First Chamber, Extended Composition) of 17 May 2018, *Bayer CropScience AG and Others v European Commission*, Cases T-429/13 and T-451/13, ECLI:EU:T:2018:280.

Judgment of the General Court (Fourth Chamber) of 14 December 2018, *Arysta LifeScience Netherlands BV, formerly Chemtura Netherlands BV v European Food Safety Authority*, T-725/15, ECLI:EU:T:2018:977.

Judgment of the General Court (Eighth Chamber) of 7 March 2019, *Antony C. Tweedale v European Food Safety Authority*, Case T-716/14, ECLI:EU:T:2019:141.

Judgment of the General Court (Eighth Chamber) of 7 March 2019, *Heidi Hautala and Others v European Food Safety Authority*, Case T-329/17, ECLI:EU:T:2019:142.

Judgment of the General Court (Second Chamber) of 21 April 2021, *Laurent Pech v Council of the European Union*, Case T-252/19, ECLI:EU:T:2021:203.

Case Law Coded in MAXQDA

Court of Justice

Judgment of the Court of 9 December 1981, *Commission of the European Communities v Italian Republic*, C-193/80, ECLI:EU:C:1981:298.

Judgment of the Court of 12 March 1987, *Commission of the European Communities v Federal Republic of Germany*, C-178/84, ECLI:EU:C:1987:126.

Judgment of the Court of 13 December 2001, *Commission of the European Communities v French Republic*, Case C-1/00, ECLI:EU:C:2001:687.

Judgment of the Court of 26 February 2002, *Commission of the European Communities v Boehringer Ingelheim Vetmedica GmbH and C. H. Boehringer Sohn*, C-32/00 P., ECLI:EU:C:2002:119.

Judgment of the Court of 20 May 2003, *Consorzio del Prosciutto di Parma and Salumificio S. Rita SpA v Asda Stores Ltd and Hygrade Foods Ltd.*, Case C-108/01, ECLI:EU:C:2003:296.

Judgment of the Court (Sixth Chamber) of 20 November 2003, *Commission of the European Communities v French Republic,* C-296/01, ECLI:EU:C:2003:626.

Judgment of the Court (Sixth Chamber) of 5 February 2004, *Commission of the European Communities v French Republic*, C-24/00, ECLI:EU:C:2004:70.

Judgment of the Court (Grand Chamber) of 12 July 2005, *The Queen, on the application of Alliance for Natural Health and Nutri-Link Ltd v Secretary of State for Health* (C-154/04) and *The Queen, on the application of National Association of Health Stores and Health Food Manufacturers Ltd v Secretary of State for Health and National Assembly for Wales* (C-155/04), Joined cases C-154/04 and C-155/04, ECLI:EU:C:2005:449.

Judgment of the Court (First Chamber) of 14 July 2005, *Commission of the European Communities v Kingdom of Spain.*Case C-135/03, ECLI:EU:C:2005:457.

Judgment of the Court (First Chamber) of 14 July 2005, *Comité Andaluz de Agricultura Ecológica v Administración General del Estado and Comité Aragonés de Agricultura Ecológica*, C-107/04, ECLI:EU:C:2005:470.

Judgment of the Court (Grand Chamber) of 6 December 2005, *The Queen, on the application of ABNA Ltd and Others v Secretary of State for Health and Food Standards Agency* (C-453/03), *Fratelli Martini & C. SpA and Cargill Srl v Ministero delle Politiche Agricole e Forestali and Others* (C-11/04), *Ferrari Mangimi Srl and Associazione nazionale tra i produttori di alimenti zootecnici (Assalzoo) v Ministero delle Politiche Agricole e Forestali and Others* (C-12/04) and *Nederlandse Vereniging Diervoederindustrie (Nevedi) v Productschap Diervoeder* (C-194/04), Joined cases C-453/03, C-11/04, C-12/04 and C-194/04, ECLI:EU:C:2005:741.

Judgment of the Court (Second Chamber) of 24 May 2007. *Maatschap Schonewille-Prins v Minister van Landbouw, Natuur en Voedselkwaliteit,* C-45/05, ECLI:EU:C:2007:296

Judgment of the Court (Grand Chamber) of 9 December 2008, *Commission of the European Communities v French Republic*, C-121/07, ECLI:EU:C:2008:695.

Judgment of the Court (First Chamber) of 5 March 2009, *Commission of the European Communities v Kingdom of Spain*, C-88/07, ECLI:EU:C:2009:123.

Judgment of the Court (Fourth Chamber) of 2 July 2009, *Bavaria NV and Bavaria Italia Srl v Bayerischer Brauerbund eV*, C-343/07, ECLI:EU:C:2009:415.

Judgment of the Court (Third Chamber) of 28 January 2010, *European Commission v French Republic*, C-333/08, ECLI:EU:C:2010:44.

Judgment of the Court (Fourth Chamber), 11 April 2013, *Karl Berger v Freistaat Bayern*, C-636/11, ECLI:EU:C:2013:227.

Judgment of the Court (Fifth Chamber), 5 November 2014, *Herbaria Kräuterparadies GmbH v Freistaat Bayern*, C-137/13, ECLI:EU:C:2014:2335.

Judgment of the Court (Second Chamber) of 16 July 2015, *ClientEarth and Pesticide Action Network Europe (PAN Europe) v European Food Safety Authority*, Case C-615/13, ECLI:EU:C:2015:489.

Judgment of the Court (Fourth Chamber) of 17 December 2015, *Société Neptune Distribution v Ministre de l'Économie et des Finances*, C-157/14, ECLI:EU:C:2015:823.

Judgment of the Court (Third Chamber) of 22 September 2016, *Breitsamer und Ulrich GmbH & Co. KG v Landeshauptstadt München*, C-113/15, ECLI:EU:C:2016:718.

Judgment of the Court (Fourth Chamber) of 19 January 2017, *Queisser Pharma GmbH & Co. KG v Bundesrepublik Deutschland*, C-282/15, ECLI:EU:C:2017:26.

Judgment of the Court (Third Chamber) of 13 September 2017, *Criminal proceedings against Giorgio Fidenato and Others*, C-111/16, ECLI:EU:C:2017:67

Judgment of the Court (Ninth Chamber) of 12 October 2017, *Kamin und Grill Shop GmbH v Zentrale zur Bekämpfung unlauteren Wettbewerbs Frankfurt am Main eV*, C-289/16, ECLI:EU:C:2017:758.

Judgment of the Court (Second Chamber) of 20 December 2017, C*omité Interprofessionnel du Vin de Champagne v Aldi Süd Dienstleistungs-GmbH & Co.OHG*, C-393/16, ECLI:EU:C:2017:991.

Judgment of the Court (Grand Chamber) of 1 October 2019, *Criminal proceedings against Mathieu Blaise and Others*, C-616/17, ECLI:EU:C:2019:800.

Judgment of the Court (First Chamber) of 3 December 2020, *Région de Bruxelles-Capitale v European Commission*, C-352/19 P, ECLI:EU:C:2020:978.

Judgment of the Court (Eighth Chamber) of 1 December 2022. *LSI - Germany GmbH v Freistaat Bayern,* C-595/21, ECLI:EU:C:2022:949.

Judgment of the Court (Fifth Chamber) of 29 June 2023, *Association interprofessionnelle des fruits et légumes frais (Interfel) v Ministre de l'Agriculture et de la Souveraineté alimentaire*, Joined Cases C-501/22 to C-504/22, ECLI:EU:C:2023:531.

Judgment of the Court (Third Chamber) of 25 April 2024. *Pesticide Action Network Europe (PAN Europe) v College voor de toelating van gewasbeschermingsmiddelen en biociden*, Joined Cases C-309/22 and C-310/22, ECLI:EU:C:2024:356.

Judgment of the Court (Third Chamber) of 25 April 2024, *Pesticide Action Network Europe (PAN Europe) v College voor de toelating van gewasbeschermingsmiddelen en biociden,* C-308/22, ECLI:EU:C:2024:350

Judgment of the Court (Eighth Chamber) of 29 July 2024, *Validity Foundation - Mental Disability Advocacy Centre v European Commission*, Case C-51/23 P, ECLI:EU:C:2024:664

Judgment of the Court (Third Chamber) of 12 September 2024, *SPAR Magyarország Kft. v Bács-Kiskun Vármegyei Kormányhivatal*, C-557/23, ECLI:EU:C:2024:737.

Judgment of the Court (Eighth Chamber) of 26 September 2024, *Verbraucherzentrale Baden-Württemberg e.V. v Aldi Süd Dienstleistungs-SE & Co. OHG,* C-330/23, ECLI:EU:C:2024:804.

Judgment of the Court (Ninth Chamber) of 4 October 2024, *UPL Europe Ltd and Indofil Industries (Netherlands) BV v European Commission*, C-262/23 P, ECLI:EU:C:2024:862.

Judgment of the Court (Second Chamber) of 4 October 2024, *Protéines France and Others v Ministre de l'Économie, des Finances et de la Souveraineté industrielle et numérique*, C-438/23, ECLI:EU:C:2024:826.

Court of First Instance/General Court

Judgement of the Court of First Instance (Third Chamber) 11 September 2002, *Pfizer Animal Health SA v Council of the European Union*, T-13/99, ECLI:EU:T:2002:209.

Judgment of the Court of First Instance (Third Chamber) of 11 September 2002, *Alpharma Inc. v Council of the European Union*, T-70/99, ECLI:EU:T:2002:210.

Judgment of the Court of First Instance (Second Chamber) of 26 February 2003, *CEVA Santé Animale SA and Pharmacia Entreprises SA v Commission of the European Communities*, Joined cases T-344/00 and T-345/00, ECLI:EU:T:2003:40.

Judgment of the Court of First Instance (Second Chamber) of 21 October 2003, *Solvay Pharmaceuticals BV v Council of the European Union*, T-392/02, ECLI:EU:T:2003:277.

Judgment of the Court (Third Chamber) of 1 April 2004, *Bellio F.lli Srl v Prefettura di Treviso*, C-286/02, ECLI:EU:C:2004:212.

Judgment of the Court of First Instance (Second Chamber) of 28 June 2005, *Industrias Químicas del Vallés, SA v Commission of the European Communities*, T-158/03, ECLI:EU:T:2005:253.

Judgment of the Court of First Instance (First Chamber) of 13 December 2006, *É. R., O. O., J. R., A. R., B. P. R. and Others v Council of the European Union and Commission of the European Communities*, T-138/03, ECLI:EU:T:2006:390.

Judgement of the Court of First Instance (First Chamber) of 13 December 2006, *Julia Abad Pérez and Others v Council of the European Union and Commission of the European Communities*, T-304/01, ECLI:EU:T:2006:389.

Judgment of the Court of First Instance (Fifth Chamber, extended composition) of 25 October 2007, *Ferriere Nord SpA v Commission of the European Communities*, T-94/03, ECLI:EU:T:2007:320.

Judgment of the Court of First Instance (Fourth Chamber) of 9 September 2008. *Bayer CropScience AG, Makhteshim-Agan Holding BV, Alfa Georgika Efodia AEVE and Aragonesas Agro, SA v Commission of the European Communities*, T-75/06, ECLI:EU:T:2008:317.

Judgment of the General Court (Third Chamber, extended composition) of 9 September 2011, *French Republic v European Commission*, T-257/07, ECLI:EU:T:2011:444.

Judgment of the General Court (Sixth Chamber) of 21 November 2013. *Günter Heede v Office for Harmonisation in the Internal Market*, T-313/11, ECLI:EU:T:2013:603.

Judgment of the General Court (Fifth Chamber), 30 April 2014, *Moritz Hagenmeyer and Andreas Hahn v European Commission*, T-17/12, ECLI:EU:T:2014:234.

Judgment of the General Court (Eighth Chamber), 13 November 2014, *Kingdom of Spain v European Commission*, T-481/11, ECLI:EU:T:2014:945.

Judgment of the General Court (Eighth Chamber) of 12 June 2015, *The Health Food Manufacturers' Association and Others v European Commission*, T-296/12, ECLI:EU:T:2015:375.

Judgment of the General Court (Eighth Chamber) of 12 June 2015, *Plantavis GmbH and NEM, Verband mittelständischer europäischer Hersteller und Distributoren von Nahrungsergänzungsmitteln & Gesundheitsprodukten eV v European Commission and European Food Safety Authority*, T-334/12, ECLI:EU:T:2015:376.

Judgment of the General Court (Fifth Chamber) of 16 March 2016, *Dextro Energy GmbH & Co. KG v European Commission*, T-100/15, ECLI:EU:T:2016:150.

Judgment of the General Court (First Chamber, Extended Composition) of 17 May 2018, *Bayer CropScience AG and Others v European Commission*, Cases T-429/13 and T-451/13, ECLI:EU:T:2018:280.

Judgment of the General Court (First Chamber, Extended Composition) of 17 May 2018, *BASF Agro BV and Others v European Commission*, T-584/13, ECLI:EU:T:2018:279.

Judgment of the Court (Grand Chamber) of 25 July 2018, *Confédération paysanne and Others v Premier ministre and Ministre de l'agriculture, de l'agroalimentaire et de la forêt*, C-528/16, ECLI:EU:C:2018:583

Judgment of the General Court (Fourth Chamber) of 14 December 2018, *Arysta LifeScience Netherlands BV, formerly Chemtura Netherlands BV v European Food Safety Authority*, T-725/15, ECLI:EU:T:2018:977.

Judgment of the General Court (Eighth Chamber) of 7 March 2019, *Antony C. Tweedale v European Food Safety Authority*, Case T-716/14, ECLI:EU:T:2019:141.

Judgment of the General Court (Fourth Chamber) of 19 September 2019, *Arysta LifeScience Netherlands BV v European Commission,* T-476/17, ECLI:EU:T:2019:618.

Judgment of the General Court (Seventh Chamber) of 9 February 2022, *Taminco BVBA and Arysta LifeScience Great Britain Ltd v European Commission*, T-740/18, ECLI:EU:T:2022:61.

Judgment of the General Court (Sixth Chamber, Extended Composition) of 14 September 2022, *Pollinis France v European Commission,* Joined Cases T-371/20 and T-554/20, ECLI:EU:T:2022:556.

Judgment of the General Court (Seventh Chamber, Extended Composition) of 4 October 2023, *Ascenza Agro, SA and Industrias Afrasa, SA v European Commission*. Case T-77/2, ECLI:EU:T:2023:602.

Judgment of the General Court (Fourth Chamber) of 21 February 2024, *Pesticide Action Network Europe (PAN Europe) v European Commission*, T-536/22, ECLI:EU:T:2024:98.

European Court of Human Rights

Leander v. Sweden, App no. 9248/81, 26 March 1987, Series A no. 116

Gaskin v. the United Kingdom, App no. 10454/83, 7 July 1989, Series A no. 160

Guerra and Others v Italy, App no. 14967/89, 19 February 1998, ECHR 1998-I 64

Oneryildiz v. Turkey, App. no. 48939/99, 30 November 2004, ECHR 2004-XII